- 中南民族大学本科教材建设项目资助
- 湖北省高等教育学会"湖北省教育科学规划2021年度专项资助重点课题：大学英语混合式教学可持续发展保障机制研究"（项目编号：2021ZA25）

# 生态环境实用英语

## Practical English for Ecological Environment

主编：罗 琪　吴晨捷　杜亚光

编者：吕康乐　杜冬云　耿 薇

　　　张大捷　李 娜　刘 潇

华中科技大学出版社
http://press.hust.edu.cn
中国·武汉

## 内 容 提 要

本书尝试以环境科学与工程学科体系作为知识框架，以学科相关重要英文表达作为教学重点，注重学科交叉融合，建设生态与环境实用英语线上课程，同时融入社会主义核心价值观，体现中华优秀传统文化，反映爱国主义精神和改革创新的时代精神，向世界展示中国环境保护成就和贡献。

全书共8章，分别从中国传统生态文化、环保问题、人口和城市化、工业和环境、自然界物质循环过程、习近平生态文明思想、农业发展和管理、碳达峰和碳中和等方面讲述各个时期人们对生态环境的认识与治理状况。

**图书在版编目(CIP)数据**

生态环境实用英语 / 罗琪，吴晨捷，杜亚光主编.—武汉：华中科技大学出版社，2024.2
ISBN 978-7-5772-0429-1

Ⅰ．①生⋯ Ⅱ．①罗⋯ ②吴⋯ ③杜⋯ Ⅲ．①生态环境－英语 Ⅳ．①X171.1

中国国家版本馆CIP数据核字(2024)第014389号

**生态环境实用英语**
Shengtai Huanjing Shiyong Yingyu

罗　琪　吴晨捷　杜亚光　主编

| | |
|---|---|
| 策划编辑：刘　平　傅　文 | |
| 责任编辑：刘　平 | |
| 封面设计：刘卿苑 | |
| 责任校对：张汇娟 | |
| 责任监印：周治超 | |
| 出版发行：华中科技大学出版社（中国·武汉） | 电话：（027）81321913 |
| 　　　　　武汉市东湖新技术开发区华工科技园 | 邮编：430223 |
| 录　　排：孙雅丽 | |
| 印　　刷：武汉开心印刷有限公司 | |
| 开　　本：787mm×1092mm　1/16 | |
| 印　　张：13.75 | |
| 字　　数：420千字 | |
| 版　　次：2024年2月第1版第1次印刷 | |
| 定　　价：48.00元 | |

本书若有印装质量问题，请向出版社营销中心调换
全国免费服务热线：400-6679-118　竭诚为您服务
版权所有　侵权必究

# 前　言

　　生态系统的可持续发展符合世界各国和各民族的利益，生态领域的全球挑战要求国际社会采取一致行动。中国将生态文明理念和生态文明建设写入宪法，纳入中国国家发展总体布局，贯彻新发展理念，坚持走生态优先、绿色低碳的发展道路，彰显了中国推动构建人类命运共同体的责任担当。

　　习近平总书记将生态文化体系建设放在首位加以强调，凸显了对其引领作用的高度重视，并明确提出了生态文明建设的时间表和近期、中期目标。第一步，确保到2035年，生态环境质量实现根本好转，美丽中国目标基本实现；第二步，到本世纪中叶，生态文明与物质文明、政治文明、精神文明、社会文明一起得到全面提升，形成绿色发展方式和生活方式，建成美丽中国。

　　在习总书记科学生态观引领下，党中央经过深思熟虑，做出了重大战略决策：我国力争2030年前实现碳达峰、2060年前实现碳中和，事关中华民族永续发展和构建人类命运共同体。联合国《生物多样性公约》缔约方大会第十五次会议（简称COP15）第一阶段会议在云南昆明召开。中国发表《中国的生物多样性保护》白皮书，介绍生物多样性保护实践和经验，为共建地球生命共同体提供中国智慧。

　　我们共享同一个地球，保护地球就是保护人类自己。"绿水青山就是金山银山。""我们必须坚持可持续发展。""中国的发展绝不会以牺牲环境为代价。"习总书记生态思想根植于中华优秀传统文化。古人如何看待人与自然的关系？"天人合一"和人与自然和谐共生理念有何联系？中国古代生态文明如何在日常生活中体现，如建筑、绘画和文学领域？大家都会担忧环境问题，诸如水污染、空气污染、气候变化、生物多样性等。这些环境问题严重程度如何？中央和地方政府采取了哪些有效措施？个人如何为环保做出贡献？欢迎来到《生态环境实用英语》探索以上问题。本教材以环保热门话题组织教学单元，深度融合五大模块——习总书记生态环保思想、中国传统生态文明、生态与环境科学系统介绍、环境保护案例、实用环保英语，旨在让学生深入了解习总书记环保思想，提升公众环境保护意识，向世界讲述中国环保故事。

为培养具有生态环保意识的国际化高水平复合型人才，本教材以环境科学与工程学科体系作为知识框架，以学科相关重要英文表达作为教学重点，注重学科交叉融合，力求满足不同层次、不同专业学生的英文表达需求。课文主要选自中国权威英文网站、政府重要文件以及习总书记生态环境保护重要论述的英文版，同时收录少量相关科普英文文章，注重趣味性以及前沿性。教材设置单元介绍、听说导入、语篇结构分析、词汇短语填空、句型操练、对话、小组口语展示活动等环节。本立体化教材与自编微课资源《绿色中国英语说》深度融合，相互连接，互为补充，为课堂教学与自主学习结合的混合式教学模式提供丰富的教学资源。

本教材由中南民族大学资源与环境学院的资深教授和外语学院的英语教学专家通力合作，精心编写而成。其中，资源与环境学院的杜冬云教授、吕康乐教授协助了编写与审阅。为保证正确的导向性，本教材课文资源主要来自China Daily网站和CGTN网站，在此向所有原作者致以诚挚的谢意。出版社的同志协助编写组安排修订日程，随时提出改进意见和建议，协助有关编写和编辑工作，为保证本教材的顺利出版付出了辛勤的劳动，在此一并致以诚挚的谢意。

主编　罗琪

2023年11月

# CONTENTS

## Unit 1  Chinese Ecological Civilization — 1

**Warm-up** — 2
**Reading Comprehension** — 3
   TEXT A — 3
   Recognizing the Importance of Stepping Up Development of an Ecological Civilization
   TEXT B — 12
   Nation's Modernization to Follow Harmonious Path
   TEXT C — 16
   The Dujiangyan Irrigation System and Qingcheng Mountain

**Supplementary Reading** — 20
   TEXT A — 20
   How Did Ancient China Protect the Environment?
   TEXT B — 22
   China's Green Philosophy Could Help Resolve Global Crises, Expert Says
   TEXT C — 23
   Rich History: A Major Draw for Visitors

## Unit 2  Common Concerns: Main Environmental Issues — 27

**Warm-up** — 28
**Reading Comprehension** — 29
   TEXT A — 29
   Scorching Heat Waves Roasting the World
   TEXT B — 37
   Second Time a Charm for Used Gadgets
   TEXT C — 44
   Climate Change Eroding Arctic Sea Ice: Report

**Supplementary Reading** — 47
   TEXT A — 47
   Air Pollution Deadlier than Thought: WHO Slashes Particle Guidelines
   TEXT B — 49

Eco-Friendly Living Becomes the Prevailing Ethos

TEXT C     51

World Must Better Protect Biodiversity

## Unit 3   Population, Urbanization and Ecology     55

### Warm-up     56

### Reading Comprehension     57

TEXT A     57

ASEAN's Rapid Urbanization Faces Sustainability Woes

TEXT B     64

A Model for Urbanization: China's City Clusters

TEXT C     69

Energy Transformation Key to Clean Heating

### Supplementary Reading     72

TEXT A     72

Hangzhou Embraces Fast Development of Digital Life

TEXT B     73

Opinion: Structural Problems Challenge China's Rapid Urbanization

TEXT C     74

China Helps Increase Global Food Security

## Unit 4   Industry and Environment     79

### Warm-up     80

### Reading Comprehension     81

TEXT A     81

COP27: Why Global Action Is Needed to Decarbonize Industries Everywhere

TEXT B     88

Road to Greater Green Consumption

TEXT C     93

Chinese Recycling Company Sees Opportunities in Circular Economy

### Supplementary Reading     95

TEXT A     95

China's Green Development in the New Era (1)

TEXT B     99

China's Green Development in the New Era (2)

TEXT C     106

Artificial Intelligence for Recycling: AMP Robotics

## Unit 5  Biogeochemical Cycles of Nature — 109

- **Warm-up** — 110
- **Reading Comprehension** — 111
  - TEXT A — 111
    The Life Cycle of Plastic Water Bottles
  - TEXT B — 118
    The Carbon Cycle
  - TEXT C — 123
    The Nitrogen Cycle in Aquariums—An Easy Guide for Beginners
- **Supplementary Reading** — 128
  - TEXT A — 128
    This Is Why Water Is Essential for Life on Earth...and Perhaps the Rest of the Universe
  - TEXT B — 129
    Seize the Moment
  - TEXT C — 132
    Improvements Being Made to Nation's Water Quality

## Unit 6  Xi Jinping Thought on Ecological Civilization — 135

- **Warm-up** — 136
- **Reading Comprehension** — 137
  - TEXT A — 137
    Can-Do Spirit
  - TEXT B — 144
    From 50 to 50
  - TEXT C — 148
    Finding Ways to Keep the Water Flowing (1)
- **Supplementary Reading** — 153
  - TEXT A — 153
    County to Boost Protection Efforts
  - TEXT B — 155
    People at the Center: From Poverty Eradication to Rural Revitalization
  - TEXT C — 158
    Finding Ways to Keep the Water Flowing (2)

## Unit 7  Agricultural Development and Management — 163

- **Warm-up** — 164

**Reading Comprehension** 164
    TEXT A 164
    Sino-German Agri-Cooperation for Mutual Benefit
    TEXT B 170
    Green Unity
    TEXT C 176
    Agriculture's Connected Future: How Technology Can Yield New Growth

**Supplementary Reading** 182
    TEXT A 182
    Should You Buy Organic? "Dirty Dozen" Pesticide List Helps You Decide
    TEXT B 183
    Biomass Technology Shows Huge Growth
    TEXT C 185
    Township Turns Waste Straw into Profit

## Unit 8  Dual Carbon Targets   189

**Warm-up** 190
**Reading Comprehension** 191
    TEXT A 191
    Ushering in the Green Transformation
    TEXT B 197
    Achieve Dual Carbon Goals in a Balanced Way
    TEXT C 201
    Climate Action to Build a Shared Future

**Supplementary Reading** 205
    TEXT A 205
    Coming Clean
    TEXT B 207
    Chinese Energy Giant Strives for Carbon Neutrality,
    Launching Mega Carbon Capture Project
    TEXT C 209
    Forest Sinks Have Critical Role in Carbon Reduction

# Unit 1

# Chinese Ecological Civilization

During the annual political sessions, environmental protection was definitely among the biggest concerns. Actually, it was also an issue that ancient Chinese paid great attention to. In fact, the world's earliest environmental protection concept, ministry and legislation were all born in China. So, how did the ancient Chinese protect the environment? Have you ever wondered how ancient Chinese viewed the relationship between man and nature? Is there any association between "Heaven and man are united as one" and the harmonious coexistence between man and nature? How was ancient Chinese ecological civilization demonstrated in various aspects of daily life, such as architecture, paintings and literature? In this unit, you will explore from the traditional Chinese wisdom that the laws of nature govern all things and that man must seek harmony with nature, to the new development philosophy emphasizing innovative, coordinated, green and open development for all. "China has always prioritized ecological progress and embedded it in every dimension and phase of its economic and social development. The goal is to seek a kind of modernization that promotes harmonious coexistence of man and nature," said President Xi Jinping.

# Practical English for Ecological Environment

## Warm-up

China, with a long history and splendid civilization, is rich in varied and profound ecological thoughts and concepts. For thousands of years, Chinese people have always longed for a harmonious coexistence of man and nature. Let's start our journey by exploring the important ancient Chinese philosophy and views related to environmental protection.

You are going to watch a video sharing Chinese wisdom. Watch it twice and complete the following two tasks. Compare your answers with your partner's.

**Task 1**  Identify the main idea.

(1) The main idea is how _____ with nature.

**Task 2**  Watch the video clip again and fill in the blanks.

(1) Noah and his family and two of every kind of animals _____ by _____.

(2) The line quoted in the video means everything _____ and when they are functioning as they should be, all things _____ _____. It stresses that if we _____, nature will _____ _____.

(3) Yu the Great _____ from his father of trying to control the flooding by _____ and _____. He _____ a way of _____ the water into the sea.

(4) The Dujiangyan irrigation system is an _____, originally _____. It uses the _____ to solve problems of _____, draining sediment, flood control and flow control.

(5) It's important to _____, with _____ of population, _____.

**Task 3**  Work in small groups. Discuss with your group members the following questions.

(1) How do you understand "Treat the earth well; it was not given to you by your parents; it was loaned to you by your children"?

(2) Do you know the differences between upper, middle and lower reaches of the Yellow River? And any measures to manage the River based on these differences?

Unit 1  Chinese Ecological Civilization

## Reading Comprehension

### TEXT A

微课

**Recognizing the Importance of Stepping Up Development of an Ecological Civilization**

Xi Jinping

1  Building an ecological civilization is vital for sustaining the development of the Chinese nation. The Chinese people have always revered and loved nature, and China's 5000-year-long civilization **embodies** a rich ecological cultural component. *The Book of Changes* states, "We look at the ornamental figures of the sky, and thereby **ascertain** the changes of the seasons. We look at the ornamental observances of society, and understand how the processes of **transformation** are **accomplished** all under heaven," and, "the ruler divides and completes the course of heaven and earth; he furthers and regulates the gifts of heaven and earth, and so aids the people." The *Dao De Jing* states, "Man takes his law from the Earth; the Earth takes its law from Heaven; Heaven takes its law from the Dao. The law of the Dao is its being what it is." The Mencius states, "If the seasons of husbandry be not **interfered** with, the grain will be more than can be eaten. If close nets are not allowed to enter the pools and ponds, the fish and turtles will be more than can be consumed. If the axes and bills enter the hill-forests only at the proper times, the wood will be more than can be used." The *Xunzi* states, "Axes must not enter the forest when the plants and trees are flourishing, lest their lives be cut short." *The Manual of Important Arts for the People*, a sixth-century agricultural encyclopedia, states, "Act according to the seasons and the nature of the land, and you will enjoy great success through little effort." These concepts all stress the importance of uniting heaven, earth, and man, following the rules of nature, and using what nature has to offer with patience and **restraint**, and show that our **ancestors** well understood the need to properly **handle** the relationship between man and nature.

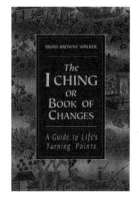

2  Environmental concepts were **elevated** to the level of state **institutions** in China at a very early time in history. A special organization **overseeing** the mountains, forests, rivers, and marshes was established, and it instituted relevant policies and decrees through the warden system.

In the *Rites of Zhou*, it is recorded that wardens were responsible for protecting the natural environment by restricting **access** to certain areas and **enforcing** relevant **prohibitions**. During the Qin (221-206 BC) and Han (206 BC-AD 220) dynasties, there were separate officers responsible for forests, rivers, shorelines, gardens, and farmlands, and the warden system in fact carried on all the way to the Qing Dynasty. Many of China's dynasties had laws for the protection of nature, and violators of these laws faced severe punishment. For example, the order issued by King Wen of Zhou (1152-1056 BC) on the attack of Chong said, "It is forbidden to destroy houses, close wells, cut trees, or disturb animals. Those who do not comply with this order shall be put to death."

3  A civilization may **thrive** if its natural surroundings thrive, and will suffer if its natural surroundings suffer. The natural environment is the basis of human survival and development, and changes to it directly **impact** the rise and fall of civilizations. The four great ancient civilizations of Egypt, Babylon, India, and China all began in regions with thick forests, **abundant** water, and **fertile** soil. The **surging** Yangtze and Yellow rivers formed the cradle of the Chinese nation, and **nurtured** our country's development into a magnificent civilization. Meanwhile, environmental **degradation**, particularly severe desertification, led to the decline of ancient Egypt and Babylon. Some areas of China also went through painful lessons in ancient times. For example, the desert sands swallowed up the once glorious and lush kingdom of Loulan. The Hexi Corridor and the Loess Plateau also once **boasted** adequate **vegetation** and water, but **excessive** deforestation to open up land for cultivation caused severe environmental damage, which in turn **aggravated** economic **decline**. The gradual shift of China's economic centers to the east and south of the country since the middle of the Tang Dynasty (618-907) was largely a result of environmental changes in western China.

ancient Egypt

ancient Babylon

ancient India

ancient China

4  On May 4, 2018, we held a ceremony marking the 200th anniversary of the birth of Karl Marx. During the ceremony, I made a point of saying that in studying Marx, we must study and practice Marxist thought on the relationship between man and nature. Marx and Engels believed that "man lives on nature," and that humans produce, live, and develop through their interactions with nature. If humans treat nature kindly, then nature will repay that kindness. However, "If man, by dint of his knowledge and inventive genius, has **subdued** the forces of nature, the latter **avenge** themselves upon him." In *Dialectics of Nature*, Engels wrote, "The people who, in Mesopotamia, Greece, Asia Minor, and elsewhere, destroyed the forests to obtain cultivable land, never dreamed that they were laying the basis for the present **devastated** condition of these

countries, by removing along with the forests the collecting centers and reservoirs of moisture. When, on the southern slopes of the mountains, the Italians of the Alps used up the pine forests so carefully cherished on the northern slopes, they had no inkling that by doing so they were cutting at the roots of the dairy industry in their region; they had still less inkling that they were thereby depriving their mountain springs of water for the greater part of the year, with the effect that these would be able to pour still more **furious** flood torrents on the plains during the rainy seasons."

**5** With history as a mirror, one can understand the rise and fall of a state. The reason why I have repeatedly emphasized the importance of taking environmental issues seriously and handling them properly is that China's environmental capacity is limited, our ecosystems are **vulnerable**, and we have still not achieved a fundamental **reversal** of environmental conditions that cause heavy pollution, significant damage, and high risk. Meanwhile, our unique geographical surroundings have exacerbated interregional imbalances. The land to the southeast of the Heihe-Tengchong Line accounts for 43% of China's total area, but is home to about 94% of its population. **Dominated** by plains, rivulets, low mountains, hills, and karst landforms, this part of China is under immense environmental pressure. The land to the northwest of the line accounts for 57% of China's total area, but is home to only about 6% of our population. Dominated by grasslands, the Gobi Desert, oases, and snowy plateaus, the ecosystems in this part of the country are extremely **fragile**. This is a very important aspect to consider when we talk about China's national conditions.

## New Words

**embody** /ɪmˈbɒdi/ *vt.*
① to express or represent an idea or a quality 表现；象征
② to include or contain something 包括；包含

**ascertain** /ˌæsəˈteɪn/ *vt.*
to find out the true or correct information about something 查明；确定

**transformation** /ˌtrænsfəˈmeɪʃn/ *n.*
a complete change in somebody/something 变革；改革

**accomplish** /əˈkʌmplɪʃ/ *vt.*
to succeed in doing or completing something 完成；实现

**interfere** /ˌɪntəˈfɪə(r)/ *vi.*
to get involved in and try to influence a situation that should not really involve you, in a way that annoys other people 干预；干扰

**restraint** /rɪˈstreɪnt/ *n.*
the act of controlling or limiting something because it is necessary or sensible to do so 抑制；管制

**ancestor** /ˈænsestə(r)/ *n.*

a person in your family who lived a long time ago 祖先；祖宗

**handle** /ˈhændl/ *vt.*

to deal with a situation, a person, an area of work or a strong emotion 处理；操作

**elevate** /ˈelɪveɪt/ *vt.*

① to give somebody/something a higher position or rank, often more important than they deserve 提升……职位

② to lift something up or put something in a higher position 提升

**institution** /ˌɪnstɪˈtjuːʃn/ *n.*

a large important organization that has a particular purpose, for example a university or bank 机构

**oversee** /ˌəʊvəˈsiː/ *vt.*

to watch somebody/something and make sure that a job or an activity is done correctly 监督；审查

**access** /ˈækses/ *n.*

the opportunity or right to use something or to see somebody/something 方法；机会

**enforce** /ɪnˈfɔːs/ *vt.*

to make sure that people obey a particular law or rule 实施；执行

**prohibition** /ˌprəʊɪˈbɪʃn/ *n.*

the act of stopping something being done or used, especially by law 禁令；禁止

**thrive** /θraɪv/ *v.*

to become, and continue to be, successful, strong, healthy, etc. 兴盛；兴隆

**impact**

/ˈɪmpækt/ *n.* the powerful effect that sth has on sb/sth 巨大影响；强大作用

/ɪmˈpækt/ *vt.* to have an effect on sb/sth 影响

**abundant** /əˈbʌndənt/ *adj.*

existing in large quantities; more than enough 大量的；充足的

**fertile** /ˈfɜːtaɪl/ *adj.*

(of land or soil) that plants grow well in 多产的；富饶的

**surge** /sɜːdʒ/ *n.*

① a sudden increase in the amount or number of something 激增

② a sudden, strong movement forward or upwards 波涛般汹涌奔腾

**nurture** /ˈnɜːtʃə(r)/ *vt.*

to care for and protect somebody/something while they are growing and developing 培养；培育

**degradation** /ˌdegrəˈdeɪʃn/ *n.*

① a situation in which somebody has lost all self-respect and the respect of other people 堕落

② the process of something being damaged or made worse 退化；下降

**boast** /bəʊst/ *vt.*

to have something that is impressive 有（引以为荣的事物）

**vegetation** /ˌvedʒəˈteɪʃn/ *n.*

plants in general, especially the plants that are found in a particular area or environment 植被

**excessive** /ɪkˈsesɪv/ *adj.*

greater than what seems reasonable or appropriate 过度的

**aggravate** /ˈæɡrəveɪt/ *vt.*

to make an illness or a bad or unpleasant situation worse 加剧；恶化

**decline** /dɪˈklaɪn/ *n.*

a continuous decrease in the number, value, quality, etc. of something 下降；衰退

**subdue** /səbˈdjuː/ *vt.*

to bring somebody/something under control, especially by using force 征服；抑制

**avenge** /əˈvendʒ/ *vt.*

to punish or hurt somebody in return for something bad or wrong that they have done to you, your family or friends 为……报仇；惩罚

**devastate** /ˈdevəsteɪt/ *vt.*

to completely destroy a place or an area 摧毁；毁灭

**furious** /ˈfjʊəriəs/ *adj.*

① very angry 愤怒的；暴怒的

② with great energy, speed or anger 强烈的；激烈的

**vulnerable** /ˈvʌlnərəbl/ *adj.*

weak and easily hurt physically or emotionally 脆弱的；敏感的

**reversal** /rɪˈvɜːsl/ *n.*

the act of changing or making something change to its opposite 撤销；推翻

**dominate** /ˈdɒmɪneɪt/ *vt.*

to control or have a lot of influence over somebody/something, especially in an unpleasant way 支配；控制

**fragile** /ˈfrædʒaɪl/ *adj.*

easily broken or damaged 易碎的

## Phrases and Expressions

**step up:** to increase the amount, speed, etc. of something 增加；加快

**carry on:** to continue moving 继续

**comply with:** to obey a rule, an order, etc.; to meet particular standards 遵守

**the rise and fall:** the vertical up and down movement of the tide 起伏；兴衰

**lead to:** to result in (something) 导致

**go through:** to continue firmly or obstinately to the end 经历

**swallow up:** to envelop or take in as if by swallowing 淹没；吞噬

**open up:** to make available or possible 开发；展现

**in turn:** in due order of succession 依次；轮流地

**make a point of:** to give one's attention to (doing something) to make sure that it happens 强调；重视

**by dint of:** by force of 凭借

**lay the basis:** to provide something (such as an idea, a principle, or a fact) from which another thing develops or can develop 奠定基础

**along with:** together with (something or someone) 连同……一起

**use up:** to exhaust of strength or useful properties 用完；用尽

**have no inkling:** to have no slight knowledge or vague notion 不知道；不了解

**take...seriously:** to treat (someone or something) as being very important and deserving attention or respect 认真对待；重视

**account for:** to be a particular amount or part of something 占

**be home to:** to be a place where something is commonly found 为……的所在地

## Exercises

**I Understanding the Text**

1. This text can be divided into four parts. The paragraph numbers of each part have been given to you in the following table. Now fill in the form with the correct choice from the following list.

① The importance of environmental protection for China and the causes

② Elevation of environmental concepts to the level of state institutions in China at a very early time in history

③ Presenting the importance of ecological civilization and rich traditional ecological culture

④ The close relationship between environmental protection and civilization and specific examples

| Parts | Paragraphs | Main ideas |
| --- | --- | --- |
| Part One | Para 1 | |
| Part Two | Para 2 | |
| Part Three | Paras 3-4 | |
| Part Four | Para 5 | |

2. Focus on the topic sentence and supporting sentences of each part and fill in the blanks with key words.

Supporting sentences **describe, explain, clarify, or give examples** of the main idea in the topic sentences. Each paragraph that you write must have enough supporting details to make the main topic clear to the reader. Likewise, a good writer makes sure that **each supporting sentence is related to the topic sentence and its controlling idea.** Good writers use many different kinds of supporting sentences and different ways of organizing them, such as quotations (Para.1), sequence of time (Para.2), places (Para.3), specific examples (Para.4) and cause and effect (Para.5).

| | Topic Sentences | Supporting Sentences |
| --- | --- | --- |
| Para.1 | Building an ecological civilization is vital for sustaining the development of the Chinese nation. | Quotations:<br>*The Book of Changes*; *Dao De Jing*; *Mencius*; *Xunzi*; *The Manual of Important Arts for the People* |
| Para.2 | | Sequence of time: |
| Para.3 | | Places: |
| Para.4 | | Specific examples: |
| Para.5 | | Causes:<br><br>Effects: |

**II Focusing on Language in Context**

1. Key Words & Expressions

A. Fill in the blanks with the words given below. Change the form where necessary. Each word can be used only once.

**accomplish interfere ancestor access thrive boast aggravated furious vulnerable dominate**

① The policy has prompted a _____ response.

② Family frictions can _____ with a child's schoolwork.

③ A business cannot _____ without good management.

④ I have _____ the task on schedule.

⑤ Attempts to restrict parking in the city centre have further _____ the problem of traffic congestion.

⑥ Migrant workers are _____ to work injury.

⑦ Chinese people use this traditional holiday to give thanks to their _____, to cherish the memory of departed relatives, to cultivate filial piety, and to reconnect with family members.

⑧ The company _____ this segment of the market.

⑨ _____ to this information is severely restricted.

⑩ Besides the vibrant cities of Siem Reap and the capital Phnom Penh, Cambodia also _____ plenty of delightful nature.

B. Fill in the blanks with the phrases given below. Change the form where necessary.

**lay the basis; comply with; step up; use up; the rise and fall; account for; go through; make a point of; open up; take seriously**

① The meeting ended with a pledge to _____ cooperation between the six states of the region.

② He's amazingly cheerful considering all he's had to _____.

③ Some beaches had failed to _____ environmental regulations.

④ People who are well are concerned with nutrition and exercise and they_____ monitoring their body's condition.

⑤ Once these plants _____ their stored reserve or tap out the underground supply, they cease growing and start to die.

⑥ I am reading a book about _____ of the Roman Empire.

⑦ Traveling to other countries _____ new possibilities and broadens one's perspective.

⑧ I believe you will _____ my advice into account.

⑨ How do you _____ the fact that unemployment is still rising?

⑩ Previous translation theories _____ on the analysis of language, that is, on the aspect of linguistics.

2. Usage

A. more than

（1）"more than＋数字"意为"超过，多于"，相当于over，此时反义短语为less than（少于，不超过）。

I have known David for more than 20 years.

我认识戴维已经有二十多年了。

注意："more than＋可数名词单数"意为"不止一个……"，在语境上虽表示复数概念，但作主语时，谓语动词用单数形式。

More than one girl holds such a view in the school.

在这所学校里，不止一个女孩持有这样的观点。

（2）"more than+ *n.*"意为"不只是，不仅仅"。

Both of us are more than workmates. We are close friends, too.

我们俩不只是同事，还是密友。

（3）"more than+ *adj./adv./v.*"意为"非常，很"，相当于very (much)。

I am more than happy to help you.

能帮助你我很高兴。

（4）"more than+句子（句中常含有can或could）"意为"非……所能……/是……难以……的"。

The beauty of the West Lake is more than I could imagine.

西湖之美是我难以想象的。

Complete the following sentences by using "more than".

① My closer friend, Li Hua, is always helping me rather than getting away from me when I am in a difficult situation that is _____（不能应对的）.

② I am _____（非常兴奋地得知）that the World Animal Protection is going to get some volunteers in China.

③ I jumped at the idea of taking the class because, after all, who doesn't want to save a few dollars? _____(不仅如此), I'd always wanted to learn chess.

④ Many thanks for you are _____（语言难以表达的）.

⑤ She told us kites were invented in China _____（2300多年前）.

B. lest

lest用于表示"以免，以防"这类含义，往往使用虚拟语气，以表达一种虚拟的情况或结果，同时告诉听者要注意避免这种可能性。主句根据实际情况使用时态，从句谓语部分用"should+动词原形"，should可省略。

Complete the following sentences by using "lest".

① He wrapped up warmly, _____.（免得感冒）

② He keeps his keys with him, _____.（以免丢失）

③ He spoke softly, _____.（以免把妹妹惊醒）

III Translation

微课

**Translate the following passage into English.**

黄河（the Yellow River）是中华民族的摇篮。她孕育了一代又一代的中国人，但她的洪水也给人们带来了灾难。勤劳的中国人民世代都在努力治理黄河水。大禹是4000多年前中国著名的治水家。大禹继承了他父亲治水的经验，经过大量调查和研究，发现了引发洪水的原因。大禹工作很努力，他致力于治水13年。在这段时间里，他没有回过家，甚至三过家门而不入。

**IV Pair Work**

Tell your classmates a story from ancient legends and myths reflecting the relationship between man and nature. Try to explore the spirits and the relationship deeply.

**TEXT B**

### Nation's Modernization to Follow Harmonious Path

Hou Liqiang

1  While simultaneously meeting people's growing needs for a better life and a beautiful environment, the Chinese path to modernization also offers other countries a new, alternative path to prosperity.

2  Qian Yong, director of the Research Center for Xi Jinping Thought on Ecological Civilization, made the remarks in an interview with *China Daily* after President Xi Jinping stressed "harmony between humanity and nature" as a major feature of China's modernization at the 20th National Congress of the Communist Party of China.

3  China's modernization features a huge population, common prosperity for all, material and cultural-ethical advancement, harmony between humanity and nature, and peaceful development. Xi, who is also general secretary of the CPC Central Committee, said at the opening session of the congress, which ran from Oct 16 to 22.

4  "Respecting, adapting to and protecting nature is essential for building China into a modern socialist country in all respects," Xi said. "We must remember to maintain harmony between humanity and nature when planning our development."

5  Qian said that under the guidance of Xi Jinping Thought on Ecological Civilization, China has rolled out a series of measures to **forge** modernization featuring harmony between humanity and nature. Xi's statement at the congress has pointed the way and provided fundamental guidance "for the country's **endeavor** to build a Beautiful China", he added.

6  "Unswervingly adhering to the new development philosophy, China has resolutely avoided seeking economic growth at the expense of the environment, and promoted transformation and upgrading of the country's industrial structure and energy mix," Qian said.

7  By mapping red lines for environmental protection, developing standards for

environmental quality and setting ceilings on resource **utilization**, the country is striving to boost economic and social development based on highly efficient utilization of its resources and green and low-carbon growth, he said.

8   According to the Ministry of Ecology and Environment, China has on average over the past 10 years managed to support annual economic growth of 6.6 percent, with a yearly increase of 3 percent in energy consumption.

9   Institutional guarantees for the construction of an ecological civilization have also been strengthened, Qian said. The country, for example, has rolled out high-profile central environmental **inspection**, which is headed by minister-level officials, to crack down on environmental **violations**. A "river chief system" that tasks heads of different levels of governments to coordinate river protection has also been established.

10   Over the past 10 years, China has **enacted** and **amended** over 30 laws and regulations that concern environmental protection, he added.

11   Qian said the Western path to modernization is not the right choice for China, not only because of the historical lessons learned from developed Western countries, but also the country's national conditions.

12   Noting that modernization is a concept shaped by Western developed nations, Qian said the Western model has seen nature **plundered** in the pursuit of profits.

13   "While creating massive material wealth, it has resulted in unimaginable damage to the environment," he said, citing major environmental disasters from the 1930s to the 1960s as examples.

14   In 1952, for instance, pollution and atmospheric conditions resulted in an incident known as the Great London Smog. Up to 4,000 people died as a direct result of the smog, according to government medical reports in the weeks following the event. Such incidents have sounded the alarm for humanity with extremely bitter prices to be paid, he said.

15   Qian said the carrying capacity of China's environment is limited and its ecosystems are fragile, however, with a population of over 1.4 billion, the nation has reversed the trend of environmental deterioration.

16   "The Western development mode of high-energy consumption and pollution is impractical if China wants to modernize the whole country," he said.

17   Qian also **underscored** the significance the Chinese modernization model has for the world.

**18**  The Chinese path has abandoned the Western development mode of "pollution first, and then treatment" and offers a new modernization path that properly handles the relationships between humanity and nature, and development and environmental protection, he said.

**19**  As China embarks on a new journey to build a modern socialist country in all respects, it will "contribute more Chinese wisdom, solutions and strength to the construction of a global ecological civilization and inject strong momentum into the endeavor of building a community with a shared future and a clean, beautiful world", Qian said.

## New Words

**forge** /fɔːdʒ/ *vt.*
to put a lot of effort into making something successful or strong so that it will last 推动
**endeavor** /ɪnˈdevə(r)/ *n.*
an attempt to do something, especially something new or difficult 努力；尽力
**utilization** /ˌjuːtəlaɪˈzeɪʃn/ *n.*
the act of using something, especially for a practical purpose 利用
**inspection** /ɪnˈspekʃn/ *n.*
an official visit to a school, factory, etc. in order to check that rules are being obeyed and that standards are acceptable 检查；视察
**violation** /ˌvaɪəˈleɪʃn/ *n.*
the act of going against or refusing to obey a law, an agreement, etc. 违反；侵害
**enact** /ɪˈnækt/ *v.*
to pass a law 制定（法律）
**amend** /əˈmend/ *v.*
to change a law, document, statement, etc. slightly in order to correct a mistake or to improve it 修改
**plunder** /ˈplʌndə(r)/ *vt.*
to steal things from a place, especially using force during a time of war 掠夺；抢劫
**underscore** /ˌʌndəˈskɔː(r)/ *vt.*
to emphasize or show that something is important or true 强调

## Phrases and Expressions

**adapt to:** to change your behaviour in order to deal more successfully with a new situation 适应
**be essential for:** completely necessary; extremely important in a particular situation or for a particular activity 对……至关重要

**at the expense of:** in a way that harms (something or someone) 以……为代价

**on average:** taking the typical example of the group under consideration 平均

**roll out:** to introduce (something, such as a new product) especially for widespread sale to the public 推出；推行

**crack down on:** to try harder to prevent an illegal activity and deal more severely with those who are caught doing it 镇压；打击

**result in:** to cause (something) to happen 导致

**sound the alarm:** to warn people 敲响警钟

**reverse the trend:** to change something completely so that it is the opposite of what it was before 扭转趋势

## Exercises

### I Comprehension Check

① What is the major feature of China's modernization, according to the interview of Qian Yong with *China Daily*?

② What is essential for building China into a modern socialist country in all respects?

③ How many laws concerning environmental protection has China enacted and amended over the past 10 years?

④ Why is the Western path to modernization not the right choice for China?

⑤ What is the result of pollution and atmospheric conditions in 1952 in London?

### II Translation

Translate into Chinese the following sentences.

① Respecting, adapting to and protecting nature is essential for building China into a modern socialist country in all respects.

② Unswervingly adhering to the new development philosophy, China has resolutely avoided seeking economic growth at the expense of the environment, and promoted transformation and upgrading of the country's industrial structure and energy mix.

③ While creating massive material wealth, it has resulted in unimaginable damage to the environment.

④ Such incidents have sounded the alarm for humanity with extremely bitter prices to be paid.

⑤ As China embarks on a new journey to build a modern socialist country in all respects, it will "contribute more Chinese wisdom, solutions and strength to the construction of a global ecological civilization".

**III Group Work**

Introduce one of the well-known international or national environmental protection institutions and organizations. You can check the official websites of these institutions for information.

**TEXT C**

## The Dujiangyan Irrigation System and Qingcheng Mountain
**Brief introduction**

1 Located on the northwest edge of the Chengdu Plain, Dujiangyan was built in the 3rd century BC and is a world famous irrigation system, controlling the waters of the Minjiang River and **distributing** it to the fertile farmland of the Chengdu Plain. Qingcheng Mountain, famous for many ancient temples, can claim to have some strong roots for China's Taoism.

2    The Dujiangyan Irrigation System, a major landmark in the development of water management and technology that is still discharging its functions perfectly, was first built in 256 BC (during the Warring States period) by magistrate Li Bing and modified and enlarged during the Tang, Song, Yuan and Ming dynasties. It is an ecological engineering **feat**, located in the western portion of the Chengdu flatlands at the junction between the Sichuan basin and the Qinghai-Tibet Plateau. Comprising two parts today—the Weir Works, located at an altitude of 726 meters, the highest point of the Chengdu plain, 1 km from Dujiangyan City, and the irrigated area, it uses natural topographic and hydrological features to solve problems of **diverting** water for irrigation, draining sediment, flood control, and flow control without the use of dams. The water from the upper valley of the Minjiang River is controlled by three key components of the Weir Works: the Yuzui Bypass Dike, the Feishayan Floodgate, and the Baopingkou Diversion Passage. These structures, with ancillary embankments and watercourses including the Baizhang Dike, the Erwang Temple Watercourse and the V-Shaped Dike, ensure a regular supply of water to the Chengdu Plain. The system has played a vital role in flood control, irrigation, water transport and general water consumption. One of the earliest irrigation systems of China and still in use today, it serves to divert waters from the Minjiang River to the West Sichuan Plain. There are many cultural relics in the neighborhood, including the Temple of the Two Kings, the Temple of the Hidden Dragon, the Bridge of Peaceful Waves and the Li Mounds.

3    Qingcheng Mountain, dominating the Chengdu Plain to the south of the Dujiangyan Irrigation System, was the birthplace of Taoism in China. With over 20 temples and religious sites for Taoism, it **exudes** a strong flavor of Taoist culture and the buildings demonstrate the Sichuan style of architecture. It is famous as the place where in 142 the philosopher Zhang Daoling founded the doctrine of Chinese Taoism. Built on the mountain during the Jin and Tang dynasties are many temples expressing Taoism culture. The mountain was regarded again as intellectual and spiritual centre of Taoism in the 17th century. The 11 important Taoist temples like the Temple of the Two Kings and the Temple of the Hidden Dragon, where Zhang Daoling preached his doctrines, display the traditional architecture of western Sichuan.

4    Qingcheng Mountain and the Dujiangyan Irrigation System were inscribed on the World Heritage List in 2000.

**Cultural heritage**

5    Large stone inscriptions by Huang Yunhu of the Qing Dynasty are prominent on the mountain, reading the Fifth Most Famous Mountain under Heaven, and the Top Peak of Qingcheng Mountain. A 2.9-meter-high and 4.5-ton statue of Li Bing, made 1,800 years ago, the first alto-rilievo stone sculpture in Chinese history, is now on display in a hall on the mountain after its excavation from a riverbed in 1974. Inscriptions recording water management methods, maps of Dujiangyan made in the Qing Dynasty and testimonials to Li Bing and his son are also on display,

side by side with precious art works by several famous modern painters such as Xu Beihong, Zhang Daqian and Guan Shanyue.

**Taoist culture**

6   Qingcheng Mountain is one of the birthplaces of Taoism. In the Eastern Han Dynasty (25-220 AD), the founder of Taoism, Celestial Master Zhang Daoling once set up his **pulpit** there to deliver lectures. During the Tang Dynasty, advocates of the newly introduced Buddhism vied with the Taoists for this base, until Emperor Xuanzong allocated the mountain to the latter.

7   As an indigenous religion of China, the Taoist religion was initiated in the Eastern Han Dynasty by Zhang Daoling and developed ever since. It is part of Taoism in the larger sense, which is deeply inspired by the theory of Laozi in the Spring and Autumn period (770-476 BC) and features the harmony of human and nature, the virtue of leisure and tranquility, and a positive attitude toward the occult and the metaphysical. The Taoist religion, which is concerned with the ritual worship of the Tao, has a profound influence upon Chinese life.

8   Qingcheng Mountain is a representative site of Taoist culture. Major religious sites here include the Natural Picture (a building complex immersed in nature), the Celestial Master's Cave, the Hall of the Ancestral Masters, Cave Facing the Sun and the Palace of Celestial Freshness.

9   All structures are shaded by dense woods and embraced by nature. The Celestial Master's Cave, perched on a cliff with only a small path leading to its entrance, houses statues of Fuxi, Shennong and Xuanyuan (three legendary primeval kings of the Han people) on its main altar.

10   The Natural Picture is a building complex made of wood. Lying in the arms of high mountains and steep cliffs, it presents a lush and primitive view of forests and pure sky. The neighboring Crane-Dwelling Village adds even more colors to the picture when the white cranes cruise gracefully among the mountain peaks.

**History**

11   During the Warring States period, people living along the banks of the Minjiang River were plagued by annual flooding. Qin governor Li Bing investigated the problem and found that the river was swelled by fast flowing spring melt-water from the local mountains that burst the banks when it reached the slow moving and heavily silted stretch below. One solution was to build a dam, but Li Bing had also been charged with keeping the waterway open for military vessels to supply troops on the frontier, so he proposed to build an artificial levee to redirect a portion of the

river's flow and then to cut a channel through Yulei Mountain to discharge the excess water upon the dry Chengdu Plain beyond.

**12** Receiving 100,000 taels of silver for the project from King Zhao of Qin, Li Bing set to work with a team said to number tens of thousands. The levee was built from long sausage-shaped baskets of woven bamboo filled with stones held in place by wooden tripods. The construction of a water-diversion levee looking like a fish's mouth took four years to finish.

**13** Cutting the channel proved to be a much more difficult problem, as the tools available to Li Bing at the time could not penetrate the hard rock of the mountain. Therefore, he used a combination of fire and water to heat and cool the rocks until they cracked and could be removed. Having worked for 8 years, a 20-metre-wide (66 ft) channel had been **gouged** through the mountain.

**14** After the system was completed, no more floods occurred. The irrigation turned Sichuan into the most productive agricultural place in China. On the east side of Dujiangyan, people built a shrine in memory of Li Bing.

## New Words

**distribute** /dɪˈstrɪbjuːt/ *v.*

If you distribute things, you hand them or deliver them to a number of people. 分发

**feat** /fiːt/ *n.*

If you refer to an action, or the result of an action, as a feat, you admire it because it is an impressive and difficult achievement. 功绩

**divert** /daɪˈvɜːt/ *v.*

To divert vehicles or travellers means to make them follow a different route or go to a different destination than they originally intended. You can also say that someone or something diverts from a particular route or to a particular place. 使改道；改道

**exude** /ɪɡˈzjuːd/ *v.*

If someone exudes a quality or feeling, or if it exudes, they show that they have it to a great extent. 充分显露；洋溢

**pulpit** /ˈpʊlpɪt/ *n.*

A pulpit is a small raised platform with a rail or barrier around it in a church, where a member of the clergy stands to speak. 布道坛

**gouge** /ɡaʊdʒ/ *v.*

If you gouge something, you make a hole or a long cut in it, usually with a pointed object. 凿

## Phrases and Expressions

**in use:** If something such as a technique, building, or machine is in use, it is used regularly by people. 在用

**on display:** put somewhere for people to see; in a display 展示

## Exercises

**Comprehension Check**

① When and where was Dujiangyan built?

_____

② Who built the Dujiangyan Irrigation System?

_____

③ What role has the system played?

_____

④ Where is the birthplace of Taoism in China?

_____

⑤ How is Taoism in the Eastern Han Dynasty?

_____

# Supplementary Reading

**TEXT A**

微课

### How Did Ancient China Protect the Environment?

1    During the annual political sessions, environmental protection was definitely among the biggest concerns. Actually, it was also an issue that ancient Chinese paid great attention to. In fact, the world's earliest environmental protection concept, ministry and legislation were all born in China. So, how did the ancient Chinese protect the environment?

**The world's earliest concept of "managing state affairs through environmental protection"**

2    In early ancient China, environmental protection was promoted to the political level. Xunzi, a famous thinker in the Warring States period,

brought up the concept of "managing state affairs through environmental protection". He stated in his book that vegetations should not be damaged at will.

3   Guan Zhong, an official 400 years ahead of Xunzi, was also an environmental protection expert. During his term of office, he claimed that "a King who cannot protect his vegetations is not qualified to be a king".

**The world's earliest "environmental protection ministry"**

4   According to a record in the Qing Dynasty, the environmental protection ministry in early ancient China was called "Yu", standing both for the institution and the official title. Although most part of its function was similar to such ministries today, its administration scope was much larger, including the mountains, forests, rivers, lakes and so on.

5   The nine ministries established by Shun, an ancient Chinese emperor, already included "Yu", the environmental protection ministry. The first "Yu" official was a man called Boyi, the earliest environmental protection minister. Record can be found in certain history documents.

6   Boyi was indeed an environmental protection expert, and a good minister. He was a capable assistant to Dayu, an ancient Chinese water-control expert. He invented wells, protecting people's drinking water from pollution. He knew a lot about animals and also called for animal protection. *The Book of Mountains and Seas* was also composed by him.

**The world's earliest "environmental protection legislation"**

7   Environmental protection legislation can be traced back to the ruling period of Dayu, which was more than 4000 years ago. During his reign, he issued a ban, forbidding people to chop wood in March or catch fish in June, the time when they were supposed to flourish.

8   In the Warring States period almost 3000 years ago, "environmental protection legislation" appeared in its true sense in Qin, which was recorded in Law of Fields and regarded as China's earliest "environmental protection legislation".

9   Many environmental protection rules can be found in Law of Fields, among which two interesting ones stood out: firstly, the river course should not be blocked; secondly, grass and trees should not be burned to be fertilizer except for summer. The last one is inspiring even for today. It can help avoid air pollution and reduce the rampant haze.

## TEXT B

### China's Green Philosophy Could Help Resolve Global Crises, Expert Says

Hou Liqiang

1  The ecological civilization China promotes that emphasizes harmony between humanity and nature will help address the prominent problems that have plagued the industrial civilization, an expert said.

2  Gong Weibin, vice-president of the National Academy of Governance, made the remarks during an online symposium themed "The study and implementation of the Xi Jinping Thought on Ecological Civilization" on Monday in Beijing.

3  Organized by the Ministry of Ecology and Environment, the event follows in the vein of President Xi's belief that "harmony between humanity and nature" must play a strong role in China's modernization, which he stated during the 20th National Congress of the Communist Party of China in October.

4  Gong said China's green modernization efforts have contributed to the development of human civilization.

5  Due to the frequency of environmental crises, Western countries have followed their own green paths, he said. While beefing up environmental governance at home, they have been exporting their industries on a large scale.

6  Though these efforts have to some extent relieved the strained relationship between humanity and nature in those countries, irreparable conflicts between rampant capital expansion and the environment persist, he said.

7  The exportation of Western countries' industries has led to the mass consumption of natural resources in developing nations, he noted. Such a development style has resulted in moral crises.

8  "The birth and establishment of every new kind of civilization must address conflicts and problems that previous kinds cannot readily resolve," he said.

9  Based on in-depth reflection on the traditional modernization methods of Western countries, the Xi Jinping Thought on Ecological Civilization has made harmony between humanity and nature an essential requirement of Chinese modernization.

10  Stressing that all members of the international community should work together to build

a sound global eco-environment, the Thought coordinates the interests of China and all other nations.

11   "It has made unique contributions in efforts to address the dilemmas and crises in mankind's development process," he said.

### TEXT C

## Rich History: A Major Draw for Visitors

Yuan Shenggao

1   A visit to Sichuan is a perfect trip back to ancient times. The province has a history of more than 3,000 years and a lot of material evidence of that history is kept intact.

2   China's history books noted the ancient Shu Kingdom in the region during the Shang Dynasty (c.16th century-11th century BC) or even earlier.

3   Many researchers doubted the records in the books until Sanxingdui Ruins was discovered in the 1920s. Massive excavations beginning in the 1950s proved this was one of China's most significant archaeological discoveries in the last century.

4   A great number of unearthed items finally consolidated historians' belief in the ancient Shu Kingdom.

5   For instance, there is a bronze mask featuring a man with protruding eyes, which is in line with the description of the kingdom's founder Can Cong.

6   The discoveries also proved that Sichuan was a cradle of Chinese civilization, with a large amount of bronze production and a high aesthetic standard in handicraft works like statues and brick drawings.

**7** What's more, the discovered items also reflect the belief and imaginations of the ancient people. For instance, a sunbird statue demonstrates their worshipping of the sun.

**8** The relics site and a museum for the relics in Deyang City are now popular destinations for many tourists interested in history and culture.

**9** The Jinsha relics site, discovered in the suburbs of Chengdu, points to a continuity of the civilization after the Sanxingdui period of the ancient Shu Kingdom.

**10** The ancient Shu Kingdom ended in 311 BC after it was conquered by the Qin State four years earlier.

**11** The rule of the Qin State marked the beginning of a new era where Sichuan was known as the "land of plenty".

**12** This was because of the Dujiangyan Irrigation System, the greatest irrigation project of its kind at the time.

**13** Qin official Li Bing and his son took charge of designing and building the project on the Minjiang River about 2,200 years ago.

**14** Since the project was completed, the Chengdu Plain has been free of flooding and people have been living affluently.

**15** Dujiangyan is the oldest surviving irrigation system in the world without a dam, and a wonder in the development of Chinese science.

**16** To this day, Dujiangyan plays a crucial role in the draining of floodwater, irrigating farms and providing water for more than 50 cities and counties across Sichuan Province.

**17** A UNESCO World Heritage site, Dujiangyan is also a popular tourist destination in the northwest of Chengdu. It is only 57 kilometers from the center of Chengdu.

**18** The provincial capital of Chengdu itself is a center of cultural tours in Sichuan.

**19** The legacies left by many historical figures living in the city have become precious resources for local tourism.

**20** Among those resources, the Temple of Marquis Wu, in memory of Zhuge Liang, the prime minister of the Shuhan State during the Three Kingdoms period (220-280). Zhuge Liang is worshipped by the people because of his wisdom and moral integrity.

**21** Sharing the same fame of the temple among tourists is the Thatched Cottage of Du Fu. Du Fu is one of the most renowned poets in the Tang Dynasty (618-907).

**22** Not far from Chengdu are the Qingcheng and Emei mountains, the hubs for Taoism and Buddhism in the region.

Unit 1 *Chinese Ecological Civilization*

## Writing

Write an introduction to one example from scenic spots, architecture, paintings or literature that reflect the concept of harmonious coexistence between man and nature. Make a presentation based on your introduction.

## Unit Project

微课

Work in groups. Talk with each other about the scenic spots, architecture, painting or literature that reflect the concepts of harmonious coexistence between human and nature.

# Unit 2

# Common Concerns: Main Environmental Issues

Environment plays an important role in healthy living and the existence of life on Planet Earth. It not only works to maintain balance in the climate and also provides all the things necessary for life. The forests, rivers, oceans and soils provide us with the food we eat, the air we breathe and the water we drink. Yet, after the beginning of human civilization and industrialization, human activities have impacted the environment in many ways leading to serious environmental issues including pollution, waste disposal, climate change, global warming, greenhouse effect, etc. Pollution refers to the release of harmful materials into the environment. The three major types of pollution are air pollution, water pollution, and land pollution. Climate change is mainly caused by greenhouse gas emission which has worsened over the last few decades with the advent of the destruction of the natural habitat, global warming, overpopulation and pollution. This chapter provides some general information on the main environmental issues we are facing and what could be done to mitigate the negative impacts.

Practical English for Ecological Environment

## Warm-up

微课

From big pieces of garbage to invisible chemicals, a wide range of pollutants ends up in our planet's lakes, rivers, streams, groundwater, and eventually the oceans. Water pollution—along with drought, inefficiency, and an exploding population—has contributed to a freshwater crisis, threatening the sources we rely on for drinking water and other critical needs. You are going to listen to an audio about water pollution. Listen to it twice and complete the following two tasks. Compare your answers with your partner's.

**Task 1** Listen to the audio and take notes according to the clues given.

(1) Causes of water pollution: natural causes, human causes
(2) Effects of water pollution
(3) Steps to control water pollution

**Task 2** Listen to the audio again and fill in the blanks with the missing information. You can refer to the notes you have taken.

(1) Water pollution occurs when _____, waste or other _____ cause the body of water to become harmful to everybody.

(2) Sometimes volcanoes, _____, _____, and silt from floods cause water pollution.

(3) Some human causes for water pollution are _____, _____ and pesticides from farms, waste from factories and _____ sites, acid rain.

(4) The effects of water pollution include: _____, _____, _____.

**Task 3** Work in small groups. Focus on one topic and discuss with your group members the following questions.

Topic 1: Water Pollution
Question 1: Why is water important to us?
Question 2: What are the possible pathways for water pollution?
Question 3: How can we protect our water resources?

Unit 2 Common Concerns: Main Environmental Issues

Topic 2: Air Pollution

Question 1: What is air? What is air pollution?

Question 2: What are the sources of air pollution?

Question 3: How can we reduce air pollution?

Topic 3: Solid Waste

Question 1: What is solid waste? Definition and classification.

Question 2: What are the adverse effects of solid waste to our environment?

Question3: How can we manage or deal with solid waste?

## Reading Comprehension

TEXT A

微课

### Scorching Heat Waves Roasting the World

Asit K. Biswas & Cecilia Tortajada

1  Global warming is producing all types of extreme weather **anomalies** around the globe. In the past, these extreme weather events were mostly reflected in major floods and **prolonged** droughts, although in the past decade, wildfires have become **rampant** and received considerable attention.

2  But only during the last few years have extreme temperatures **drawn** global attention, not least because they have been occurring all over the world and all too frequently. This is not to say **intense** heat waves were not observed in the post-2000 period before the past few years. In August 2003, some 72,000 people died in Europe because of an extreme heat wave considered the hottest in 500 years. Similarly, in 2010, some 56,000 Russians died in another extreme heat wave.

**Heat waves intensifying around the world**

3  According to the United Nations Office for Disaster Risk Reduction (UNDRR), the world experienced 405 extreme heat waves from 1998 to 2017, which affected some 97 million people, claimed 166,346 lives and caused economic losses of $61 billion.

29

4   The UNDRR reported economic losses from all types of natural disasters during those 20 years. For example, storm damage accounted for some 55 percent of the incidents, compared with about 11 percent for extreme temperature events.

5   There were also serious flaws in reporting about the natural disasters across the globe. To give an example, less than 14 percent of all natural disasters in Africa were reported. That means the UNDRR data on the impacts of extreme temperature events and other types of natural disasters were grossly underestimated.

6   Unlike earthquakes and hurricanes, other extreme weather events occur across the world. Asia, the most populous continent, <u>bore the brunt of</u> global natural disasters, including intense heat waves, accounting for 69 percent of all deaths and 78 percent of the total economic loss from 1998 to 2017. Also, 85 percent of all the affected were in Asia.

**Developing states suffer more from heat waves**

7   Heat waves, or extreme temperatures, are defined based on a country's temperature profile. For instance, the Netherlands defines it as at least five consecutive days of the maximum temperature exceeding 25 degrees Celsius, with temperatures exceeding 30 C for at least three days. In the United States, the **criterion** is 90 degrees Fahrenheit (32.5 C) for three or more consecutive days.

8   For residents of cities such as Beijing, New Delhi, Cairo and São Paulo, temperatures of 30 C or 32.5 C—considered very high in the Netherlands and the US—are normal in early summer itself. What is considered a heat wave in Western countries is often below average summer temperature in most of the developing countries. Extreme heat waves therefore <u>have a more devastating impact</u> on developing countries, almost all of which are in the **tropics** and semi-tropics. In contrast, most of the developed countries are in the temperate region.

9   However, as the world warms, even countries in the temperate region such as Canada and the United Kingdom are facing extreme heat waves, and their maximum temperature records, set in the past, are falling. Take Canada for example. Between June 25 and July 1, 2021, temperatures in some areas of British Columbia reached 49.6 C, higher than that in Kuwait or Dubai. Similarly, the temperature in London reached the historical high of 40.2 C last month.

10   What is really worrying is that previous <u>records were broken</u> by <u>a fraction of</u> 1 C. Now, the increase is by 3-5 C, as seen in Canada and the UK.

11   For a city like New Delhi, the situation is even worse. In May 2022, it recorded its highest ever temperature of 49 C, beating the record of 45.6 C set on April 29, 1941. China, too, is experiencing intense heat waves. In the previous two months alone, heat waves/high temperatures affected about 900 million people in the country.

12   According to China's National Climate Center, 71 national weather stations reported record high temperatures this year by the middle of July. While Xi'an, Shaanxi Province, witnessed 20 days of 35 C-plus temperatures in June, Shanghai recorded 40.9 C in July, matching its highest ever recorded temperature. Moreover, a scorching heat wave is forecast to **smother** most parts of the country in the coming two weeks. The summer of 2022 is on track to become the hottest since China began **compiling** complete meteorological records in 1961.

13   Even the Arctic town of Verkhoyansk in Siberia, Russia, registered 38 C in June 2020.

14   No part of the world is free from extreme heat.

**Devastating impacts of heat waves**

15   There are many direct and indirect impacts of heat waves. Many of them are known; others are not. And since elderly people are the most **vulnerable** to heat waves and heat-strokes, they need to be especially protected against high temperatures.

16   Yet heat waves not only **jeopardize** human health, they also create other serious societal and economic hazards. For instance, they can reduce crop output by 20-30 percent, if not altogether destroy the crops, **triggering** food price hikes which affect the poor the most.

17   Similarly, higher temperatures **invariably** mean increased use of air-conditioners, which in turn requires more electricity to run. And higher energy consumption generally means an increase in greenhouse gas emissions, which in turn further intensifies global warming and raises temperatures. A lose-lose situation, indeed.

18   Heat waves also act as a kind of atmospheric lid which **traps** air pollutants. Nearly 99 percent of the global population is already breathing air that does not meet the World Health Organization's health standards, and with heat domes worsening the air quality, the health risks and death rates will rise.

19   Many indirect impacts of heat waves are not fully understood. Long periods of dry and hot weather or high temperatures are likely to make tree roots grow deeper underground to seek water, and thus reduce soil moisture deep below the ground surface, causing subsidence. In fact, studies show that more than 7.65 million houses in the UK could become vulnerable to medium-to-high **subsidence** by the 2080s. During the record heat wave in the UK in 2018, about 10,000 households sought insurance claims of about £64 million for soil subsidence in just three months. Such claims are likely to increase exponentially.

## What can be done to curb heat waves?

**20**  Unfortunately, not much research has been done on how to keep urban centers cool amid heat waves until recently. But since we know that cities are 2-4 C warmer than their surrounding areas, there is a need to <u>carry out</u> studies to find ways to protect cities against "urban heat island effect". Buildings, roads and other infrastructure absorb the sun's rays and re-emit the heat more than natural landscapes like a cluster of trees, forests and water bodies, so urban areas where such structures are concentrated and greenery is **sparse** become "islands" of heat or higher temperatures <u>relative to</u> the outlying areas.

**21**  Actually, research is accelerating on how buildings and cities can be designed and landscaped to keep them cooler. Even the buildings' colors could be a factor, as lighter colours reflect more of the sun's rays and can help to lower the temperatures in the buildings' interiors and the areas around them.

**22**  The speed at which global warming is intensifying, global temperatures have already risen by over 1 C, and the frequency and intensity with which it's causing extreme weather events, including heat waves, has taken most scientists by surprise.

**23**  Climate change will continue to worsen till at least 2060 given the **cumulative** global GHG emissions since the Industrial Revolution, irrespective of what happens in the coming years. Accordingly, people all over the world will have to <u>live with</u> and <u>adapt to</u> extreme weather conditions and events for many decades to come.

### New Words

**scorching** /ˈskɔːtʃɪŋ/ *adj.*
very hot 酷热的
**roast** /rəʊst/ *v.*
to become or to make sth become very hot in the sun or by a fire 曝晒；烘烤
**anomaly** /əˈnɒməli/ *n.*
a thing, situation, etc. that is different from what is normal or expected 异常事物；反常现象
**prolong** /prəˈlɒŋ/ *v.*
to make sth last longer 延长
**rampant** /ˈræmpənt/ *adj.*
existing or spreading everywhere in a way that cannot be controlled 泛滥的；猖獗的

**intense** /ɪnˈtens/ *adj.*

very great; very strong 很大的；十分强烈的

**criterion** /kraɪˈtɪəriən/ *n.*

a standard or principle by which sth is judged, or with the help of which a decision is made （评判或做决定的）标准，准则

**tropics** /ˈtrɒpɪks/ *n.*

the area between the two tropics, which is the hottest part of the world 热带；热带地区

**smother** /ˈsmʌðə(r)/ *v.*

to cover sth/sb thickly or with too much of sth （用某物）厚厚地覆盖

**compile** /kəmˈpaɪl/ *v.*

to produce a book, list, report, etc. by bringing together different items, articles, songs, etc. 编写（书、列表、报告等）；编纂

**vulnerable** /ˈvʌlnərəbl/ *adj.*

weak and easily hurt physically or emotionally （身体上或感情上）脆弱的，易受……伤害的

**jeopardize** /ˈdʒepədaɪz/ *v.*

to risk harming or destroying sth/sb 冒……的危险；损害

**trigger** /ˈtrɪɡə(r)/ *v.*

to make sth happen suddenly 发动；引起

**invariably** /ɪnˈveəriəbli/ *adv.*

always 始终如一地；一贯地

**trap** /træp/ *v.*

to catch or keep sth in a place and prevent it from escaping, especially so that you can use it 收集；吸收

**subsidence** /səbˈsaɪdns; ˈsʌbsɪdns/ *n.*

the process by which an area of land sinks to a lower level than normal, or by which a building sinks into the ground （地面或建筑物的）下沉，沉降，下陷

**sparse** /spɑːs/ *adj.*

only present in small amounts or numbers and often spread over a large area 稀少的；稀疏的

**cumulative** /ˈkjuːmjələtɪv/ *adj.*

having a result that increases in strength or importance each time more of sth is added （在力量或重要性方面）聚积的，积累的

## Phrases and Expressions

**receive attention**：If you receive attention, you get someone's attention or interest. 获得

关注；受到关注

**not least:** especially or particularly 尤其是；特别是

**claim sb's lives:** If a violent event, fighting, or disease claims someone's life, it kills that person. 夺走……的生命

**bear the brunt of :** to bear the brunt or take the brunt of something unpleasant means to suffer the main part or force of it 首当其冲

**have an impact on:** to have a strong effect or influence on a situation or person 对……产生影响

**break the record:** to do something better than the best known speed, time, number, etc. previously achieved 打破纪录

**a fraction of:** a small part or amount of sth 小部分；少量

**be on track:** to be doing the right thing in order to achieve a particular result 步入正轨；做法对头

**be free from:** lacking; without 免于；免除

**carry out:** to put into execution 实施

**relative to:** in connection with 相较于

**live with:** to accept a difficult situation that is likely to continue for a long time 忍受

**adapt to:** to change your behaviour in order to deal more successfully with a new situation 适应（新情况）

## Exercises

**I Understanding the Text**

1. Recognize the sequence in which the passage is organized. Fill in the form with the correct choice form the following list.

① Developing states suffer more from heat waves.

② Heat weaves have been intensifying around the world.

③ There are many direct and indirect impacts of heat waves.

④ The extreme weather events have drawn global attention.

⑤ There is a need to carry out studies to find ways to curb heat waves.

| Parts | Paragraphs | Main ideas |
| --- | --- | --- |
| Part One | Paras.1-2 | |
| Part Two | Paras.3-6 | |
| Part Three | Paras.7-14 | |
| Part Four | Paras.15-19 | |
| Part Five | Paras.20-23 | |

2. Focus on the sentences in the text that provide specific examples and give comparisons and contrasts. Fill in the blanks with key information and pay attention to the markers of exemplification and comparisons and contrasts.

① _____, storm damage accounted for some _____ of the incidents, _____ for extreme temperature events.

② _____, _____ of all natural disasters in Africa were reported.

③ _____, other extreme weather events occur across the world.

④ _____, the Netherlands defines it as at least five consecutive days of the maximum temperature _____ Celsius, with temperatures exceeding 30 C for at least three days.

⑤ Extreme heat waves therefore have a more devastating impact on developing countries, almost all of which are _____. _____, most of the developed countries are _____.

⑥ However, as the world warms, even countries in the temperate region such as _____ are facing extreme heat waves, and their maximum temperature records, set in the past, are falling.

⑦ _____. Between June 25 and July 1, 2021, temperatures in some areas of British Columbia reached 49.6 C, higher than that in Kuwait or Dubai.

⑧ _____, the temperature in London reached _____ last month.

⑨ What is really worrying is that _____. _____, the increase is _____, as seen in Canada and the UK.

**II Focusing on Language in Context**

1. Key Words & Expressions

A. Fill in the blanks with the words given below. Change the form where necessary. Each word can be used only once.

**roast   prolong   anomaly   rampant   smother   compile   vulnerable   jeopardize   trigger   invariably**

① Nuts can _____ off a violent allergic reaction.

② In cases of food poisoning, young children are especially _____.

③ Once the shrubs begin to _____ the little plants, we have to move them.

④ Unemployment is now _____ in most of Europe.

⑤ We are trying to _____ a list of suitable people for the job.

⑥ The fire was hot enough to _____ everyone who stood close to it.

⑦ The operation could _____ his life by two or three years.

⑧ The _____ of the social security system is that you sometimes have more money without a job.

⑨ We lay on the beach and _____ in the Mediterranean sun.

⑩ His intuition is _____ correct.

## Practical English for Ecological Environment

B. Fill in the blanks with the phrases given below. Change the form where necessary.

**receive attention; not least; claim...lives; bear the brunt of; a fraction of; be free from; carry out; live with; adapt to; relative to**

① Our policies should serve the common good and must _____ the influence of vested interests.

② The documentary caused a lot of bad feeling, _____ among the workers whose lives it described.

③ This represents only _____ the couples who want a divorce.

④ Now, the employment of college students began to _____.

⑤ Much recent work has examined the claim that women encounter increasing obstacles _____ men as they move up the organizational ladder in business.

⑥ Police say they believe the attacks were _____ by nationalists.

⑦ They learned to _____ each other's imperfections.

⑧ The war has _____ thousands of _____.

⑨ It took him a while to _____ himself _____ his new surroundings.

⑩ Young people are _____ unemployment.

2. Usage

A. Complete the following sentences in the text by using the comparative and superlative degrees.

① Extreme heat waves therefore have a _____ impact on developing countries, almost all of which are in the tropics and semi-tropics. (devastating)

② Between June 25 and July 1, 2021, temperatures in some areas of British Columbia reached 49.6 C, _____ in Kuwait or Dubai. (high)

③ For a city like New Delhi, the situation is even _____. (bad)

④ The summer of 2022 is on track to become _____ since China began compiling complete meteorological records in 1961. (hot)

⑤ And since elderly people are _____ to heat waves and heat-strokes, they need to be especially protected against high temperatures. (vulnerable)

⑥ For instance, they can reduce crop output by 20-30 percent, if not altogether destroy the crops, triggering food price hikes which affect the poor _____. (much)

B. Complete the following sentences in the text by using attributive clauses led by "which".

① According to the United Nations Office for Disaster Risk Reduction (UNDRR), the world experienced 405 extreme heat waves from 1998 to 2017, _____. (affected some 97 million people)

② Extreme heat waves therefore have a more devastating impact on developing countries, almost _____. (all of them are in the tropics and semi-tropics)

③ For instance, they can reduce crop output by 20-30 percent, if not altogether destroy the crops, triggering food price hikes _____. (affect the poor the most)

④ And higher energy consumption generally means an increase in greenhouse gas emissions, _____. (in turn further intensifies global warming and raises temperatures)

⑤ The speed _____, global temperatures have already risen by over 1 C, and the frequency and intensity _____, including heat waves, has taken most scientists by surprise. (global warming is intensifying; it's causing extreme weather events)

**III Translation**

Translate the following passage into English.

随着世界人口越来越密集（densely populated），空气污染已经成了严重的问题。空气污染主要来源于四个主要的人类活动领域：工业、能源业、交通运输业以及农业。经营工厂，为火车、飞机和公共汽车提供动力都需要能源，几乎所有这些能源都是通过燃烧燃料产生的，这就会造成空气污染。科学家正在研究能减少环境破坏的新发电方式。人们的公共环保意识增强，回收利用（recycling）等活动开始出现。

**IV Pair Work**

Discuss how climate change affected your daily life and what can be done to alleviate the adverse effects.

**TEXT B**

微课

## Second Time a Charm for Used Gadgets

Fan Feifei

1   Ever wondered what happens to hundreds of millions of unwanted smartphones, laptops and other **discarded** electronic devices as consumers switch to updated gadgets every once in a while?

2   Nowadays, such rejected devices are finding their way to online platforms, which either resell them as secondhand goods at attractive prices or channel them to recycling firms if they are deemed low-value items.

3   An increasing number of young internet-**savvy** shoppers, who are also environmentally conscious, are doing away with old habits of rejecting pre-owned goods

and instead are looking for high-quality and affordable products on secondhand trading platforms.

4   China remains the world's largest smartphone market. According to a report by the China Association of Circular Economy, in the past five years, 430 million mobile phones have been sold annually on average in China, accounting for 30 percent of global smartphone sales.

5   Meanwhile, about 370 million handsets were discarded in 2020, and the figure rose to 404 million last year, the report said. However, only 2 percent of them have been recycled using regular channels, said a report in August, **citing** data from the Ministry of Industry and Information Technology.

6   <u>Compared with street **vendors** who engage in mobile phone recycling, online platforms provide users with a more reliable and stable recycling channel with transparent pricing structures.</u>

7   Shanghai-based electronics recycling platform ATRenew Inc., formerly known as Aihuishou— "love of recycling"—focuses on the trading and recycling of secondhand electronic products such as cameras, phones and laptops, with an aim to minimize the negative impact of discarded consumer electronics on the environment.

8   Established in 2011, ATRenew, which stands for "all things renew", has opened 1,629 brick-and-mortar stores in 241 cities across the nation as of June 30.

9   The company reported its total net revenue grew 14.9 percent year-on-year to 2.15 billion yuan ($298.6 million) in the second quarter, while its gross merchandise volume or GMV reached 8.6 billion yuan, up 10.3 percent.

10   <u>"Although our business is facing short-term headwinds from the COVID-19 pandemic, we firmly believe that the demand for electronic device recycling, trade-ins and other value-added services will grow along with the long-term development of the circular economy in China,"</u> said Chen Xuefeng, co-founder, chairman and CEO of ATRenew.

11   The company now operates four business lines, covering consumer-to-business or C2B electronics recycling platform Aihuishou, B2B used electronics trading platform Paijitang, B2C secondhand platform Paipai as well as AHS Device, which concentrates on electronics recycling in the global market.

12   Recycled mobile phones firstly undergo irreversible information and data removal. Lower-end phones will be <u>handed over</u> to electronic waste recycling companies for **dismantling** and metal extraction, while premium castaways will be resold as used goods after professional

processing.

**13**　Du Xiaochen, vice-president of ATRenew, said they are <u>ratcheting up</u> efforts to build online and offline recycling systems and boost supply chain capacity while reducing costs and improving transaction efficiency.

**14**　Du said the company is seeking new sources of revenue by offering smartphone maintenance, peripheral product sales and other value-added services, as well as expanding product categories covering luxury products and photographic equipment at its offline stores in first and second-tier cities.

**15**　It has collaborated with smartphone makers to launch the "trading in an old one for a new one" service, established standardized quality inspection systems, and improved after-sales services for secondhand electronic devices.

**16**　In 2019, the company <u>merged with</u> Paipai, a secondhand trading platform backed by e-commerce **behemoth** JD, <u>in a bid to</u> further standardize the electronics recycling process and improve efficiency by **leveraging** JD's retail, logistical and technological strengths.

**17**　Li Zhifeng, a 27-year-old software engineer from Taiyuan, capital of Shanxi Province, is considering upgrading to the newly launched iPhone 14 from the iPhone X, which he bought about four years ago.

**18**　"I plan to sell my old phone. However, I don't trust street vendors who engage in mobile phone recycling as I fear my personal information might be leaked and the value of the used phone can't be fairly assessed," Li said.

**19**　Li added he would probably <u>opt for</u> a formal channel that offers a clear and reasonable price for secondhand commodities.

**20**　The market scale of China's secondhand consumer electronic products is expected to amount to 967.3 billion yuan in 2025 from 252.2 billion yuan in 2020, with a compound annual growth rate reaching 30.8 percent during this period, according to consultancy CIC.

**21**　Pan Helin, co-director of the Digital Economy and Financial Innovation Research Center at Zhejiang University's International Business School, called for efforts to **bolster** the standardized and regulated development of the smartphone recycling industry, and establish relevant industry standards covering the removal of personal information as well as market-based price setting for recycled phones.

22    Pan said electronics recycling platforms should strengthen cooperation with existing smartphone makers to further reduce transaction costs, and stimulate people's willingness to sell phones they no longer use.

23    Last July, the National Development and Reform Commission released a development plan to **spur** the circular economy during the 14th Five-Year Plan period (2021-2025). Circular economy refers to a model focusing on recycling and reusing materials and resources.

24    The NDRC has <u>called for</u> efforts to accelerate the establishment of a recycling system for waste goods and materials, improve the discarded goods recycling network, enhance the processing and utilization of renewable resources, and develop the secondhand goods trading and refurbishing industry in an orderly manner.

25    <u>Developing the circular economy is conducive to helping achieve the country's goal of peaking carbon emissions by 2030 and realizing carbon neutrality by 2060</u>, experts said.

26    According to a report released by global consultancy Frost & Sullivan and the Institute of Energy, Environment and Economy at Tsinghua University, the trading of underutilized or unwanted goods—which serves as an important part of a circular economy—can effectively promote the efficient utilization of recyclable resources in China and achieve carbon emissions reduction.

27    The report said each transaction of an unwanted mobile phone can reduce at least 25 kilograms of carbon emissions, while trading a discarded refrigerator can realize 130 kg of such emission reductions.

28    "<u>The recycling of smartphones via regular channels is conducive to prolonging the life cycle of mobile phones, reducing carbon emissions and contributing to the sustainable development of the environment</u>," said Wu Shenkuo, law professor and assistant dean of the Internet Development Research Institute at Beijing Normal University.

29    More efforts should be made to select formal channels and reliable service providers to finish the recycling process, especially ensuring the full erasure of personal data, Wu said, adding that authorities should strengthen supervision and management to protect user privacy.

30    Zhuanzhuan, another online trading platform for used goods in China, has set up operation centers and stores in Beijing; Shenzhen, Guangdong Province; Qingdao, Shandong Province, and Chengdu, Sichuan Province, to promote the recycling and circulation of idle resources, including used mobile phones. Its intelligent sorting,

warehousing and quality inspection center for different types of secondhand goods in Qingdao went into operation in July.

31  Based on massive transaction data and artificial intelligence-powered algorithms, Zhuanzhuan has launched a secondhand commodity pricing guidance system, which can accurately match user needs and enhance the willingness of consumers to trade their unwanted goods on the platform.

32  Low-carbon, sustainable development has been an irresistible trend driving social and economic development, said Huang Wei, CEO of Zhuanzhuan. "We have seen that people's consumption concepts have changed significantly, with their demand becoming more and more diversified, as they are willing to buy and use secondhand goods and sell their unused and unwanted items."

33  Huang said its core business—the recycling of consumer electronics products—registered triple-digit growth in 2021, while noting the circulation and recycling of unused commodities, including secondhand smartphones, can effectively reduce carbon emissions.

34  "In the past six years, we have achieved more than 1.78 million metric tons of carbon emission reductions by working with users to bolster the circulation of unwanted goods, which is equivalent to reducing the energy consumption of fossil fuel-powered vehicles having traveled about 10 billion kilometers," Huang said.

35  Huang said the company will increase investment in AI-powered intelligent appraisals, dynamic pricing systems and other scientific and technological innovations, build digitalized platforms, and notably improve the circulation efficiency of secondhand goods so as to meet people's diversified consumption demand.

36  Mo Daiqing, a senior analyst at the Internet Economy Institute, a domestic consultancy, said the country's secondhand market, which is still at a nascent stage, has huge growth potential, and the penetration rate of online recycling and trading platforms is still relatively low.

37  Meanwhile, the new business model has encountered many problems. Mo said sellers have more information than buyers before transactions occur on these platforms, which may result in the appearance of some fake and low-quality products and the occurrence of fraudulent behavior during transaction processes.

38  "How to deal with the above issues and establish mutual trust between sellers and buyers is key to the secondhand market," Mo added.

## New Words

**gadget** /ˈgædʒɪt/ *n.*
a small tool or device that does sth useful 小器具；小装置

**discard** /dɪˈskɑːd/ *v.*
to get rid of sth that you no longer want or need 丢弃；抛弃

**savvy** /ˈsævi/ *adj.*
having practical knowledge and understanding of sth; having common sense 有见识的；懂实际知识的

**cite** /saɪt/ *v.*
to mention sth as a reason or an example, or in order to support what you are saying 提及（原因）；举出（示例）

**vendor** /ˈvendə(r)/ *n.*
a person who sells things, for example food or newspapers, usually outside on the street 小贩；摊贩

**dismantle** /dɪsˈmæntəl/ *v.*
If you dismantle a machine or structure, you carefully separate it into its different parts. 拆除

**behemoth** /bɪˈhiːmɒθ/ *n.*
a very big and powerful company or organization 巨头（指规模庞大、实力雄厚的公司或机构）

**leverage** /ˈliːvərɪdʒ/ *v.*
use something to maximum advantage 最大限度的利用；最优化使用

**bolster** /ˈbəʊlstə(r)/ *v.*
to improve sth or make it stronger 改善；加强

**spur** /spɜː(r)/ *v.*
to make sth happen faster or sooner 促进，加速

## Phrases and Expressions

**switch to:** to begin doing, using, consuming, etc., something new or different 转变；转到

**find one's way to:** to look for and find where one needs to go in order to get somewhere 找到；到达

**do away with:** to get rid of something or stop using something 摆脱

**engage in:** to take part in sth; to make sb take part in sth （使）从事，参加

**face a headwind:** to fave a force or influence that inhibits progress

**hand over:** If you hand something over to someone, you give them the responsibility for dealing with a particular situation or problem. （把某事）交给……负责

**ratchet up:** If something ratchets up or is ratcheted up, it increases by a fixed amount or degree, and seems unlikely to decrease again. 稳步提高；使……稳步提高

**merge with:** to combine or make two or more things combine to form a single thing （使）合并，结合，并入

**in a bid to: in an effort to do sth or to obtain sth** 为了；试图

**opt for:** to choose to take or not to take a particular course of action 选择；挑选

**call for:** If you call for something, you demand that it should happen. 要求

**be conducive to:** making it easy, possible or likely for sth to happen 使容易（或有可能）发生的

## Exercises

### I Comprehension Check

① Which country remains the world's largest smartphone market?

② What is the focus of ATRenew Inc.?

③ What happens to recycled mobile phones?

④ According to Pan Helin, what should recycling platforms do to further reduce transaction costs?

⑤ According to Huang Wei, what has been an irresistible trend driving social and economic development?

### II Translation

Translate into Chinese the underlined sentences in the article.

① Nowadays, such rejected devices are finding their way to online platforms, which either resell them as secondhand goods at attractive prices or channel them to recycling firms if they are deemed low-value items.

② Compared with street vendors who engage in mobile phone recycling, online platforms provide users with a more reliable and stable recycling channel with transparent pricing structures.

③ Although our business is facing short-term headwinds from the COVID-19 pandemic, we firmly believe that the demand for electronic device recycling, trade-ins and other value-added services will grow along with the long-term development of the circular economy in China.

④ Developing the circular economy is conducive to helping achieve the country's goal of peaking carbon emissions by 2030 and realizing carbon neutrality by 2060.

⑤ The recycling of smartphones via regular channels is conducive to prolonging the life cycle of mobile phones, reducing carbon emissions and contributing to the sustainable development of the environment.

**III Group Work**

What are the benefits of garbage sorting? China has made great progress in garbage sorting. Illustrate the process with examples in your hometown.

**TEXT C**

## Climate Change Eroding Arctic Sea Ice: Report

Zhang Zhihao

1  The Arctic region lost over 2 million square kilometers of sea ice from 2002 to 2021, leaving around 3.96 million sq km of sea ice as of September 2021, according to a report by the National Remote Sensing Center of China, an institution **affiliated** with the Ministry of Science and Technology.

2  The Arctic, one of Earth's key ecosystems and home to over 4 million people, is experiencing rising temperatures, accelerating ice melt and increased vegetation growth due to climate change, the report found.

3  From 2002 to 2021, multilayer ice—sea ice that persists in the ocean from year to year—

was disappearing at a rate of 127,000 sq km per year. Although this type of ice typically reappears in the winter, the overall coverage of multilayer ice has been decreasing at a rapid rate.

4  The consequences of melting sea ice are complex. On one hand, sailing through Arctic shipping channels has become more convenient due to less ice blocking key waterways, but on the other hand, the habitats for many Arctic animals such as polar bears and seals are being threatened.

5  Zhao Jing, director of the National Remote Sensing Center of China, said the report not only showcases China's remote-sensing technologies, but also represents the country's contribution to the international scientific community on tackling climate change.

6  Zhou Chenghu, a noted geologist and an academician of the Chinese Academy of Sciences, said the impact of climate change on the polar regions is closely related to human life, especially those living in coastal regions.

7  "Environmental changes in the Arctic region are a key indicator of climate change, and this is why the international scientific community is paying close attention to it," he said.

8  In November, an international report said Arctic sea ice is melting at increasing rates far faster than scientists had expected, and summertime Arctic sea ice could vanish by 2050.

9  Zhou said scientists are still debating when summertime Arctic sea ice would be gone, but it is clear that the international community should be **galvanized** to safeguard this distinct environment to prevent climate catastrophes.

10  Huang Huabing, a professor of remote sensing at Sun Yat-sen University, said the Arctic region has been a focus of climate research due to the effect of Arctic amplification, a complex phenomenon that saw the region warming over twice as fast as the rest of the planet.

11  Due to rising temperatures, from 2002 to 2021, about 77.4 percent of the Arctic's land surface saw an increase of vegetation coverage, totaling nearly 5.5 million sq km, or the entire area of the Amazon rainforest, he said.

12  However, the greening of the Arctic region is extremely **fragile** and heavily influenced by temperature, seasonal snowfall and human activities. In areas where there are many human activities, some of the newly emerged vegetation has already been damaged.

13  Liu Zhichun, the deputy director of the National Remote Sensing Center of China, said climate change in the Arctic region would affect the entire oceanic ecosystem.

**14** In order to achieve the United Nations' Sustainable Development Goal 13 on tackling climate change, Liu said it is **vital** that scientists establish a **robust** and systematic mechanism to monitor and preserve the Arctic region.

### New Words

**erode** /ɪˈrəʊd/ *v.*
to gradually destroy the surface of sth through the action of wind, rain, etc.; to be gradually destroyed in this way 侵蚀；腐蚀

**affiliate** /əˈfɪlieɪt/ *v.*
to link a group, a company, or an organization very closely with another larger one 使隶属，使并入（较大的团体、公司、组织）

**galvanize** /ˈɡælvənaɪz/ *v.*
to make sb take action by shocking them or by making them excited 使振奋；激励

**fragile** /ˈfrædʒaɪl/ *adj.*
weak and uncertain; easily destroyed or spoilt 不牢固的；脆弱的

**vital** /ˈvaɪtl/ *adj.*
necessary or essential in order for sth to succeed or exist 必不可少的；对……极重要的

**robust** /rəʊˈbʌst/ *adj.*
strong and not likely to fail or become weak 强劲的；富有活力的

### Exercises

**Comprehension Check**

① How much sea ice has been lost from 2002 to 2021?

② According to the report, what will the Arctic see in the future because of climate change?

③ Why are the consequences of melting sea ice complex?

④ Why has the Arctic region been a focus of climate research?

⑤ What factors heavily influence the greening of the Arctic region?

## Supplementary Reading

### TEXT A

**Air Pollution Deadlier than Thought: WHO Slashes Particle Guidelines**

Giulia Carbonaro

1   The World Health Organization (WHO) has halved its recommended level of tiny particles from burning fossil fuels allowed in the atmosphere, after new evidence has proved air pollution is deadlier than previously thought.

2   It's the first update to the WHO's guidelines on global air pollution in 16 years, after data collected since 2005 has shown bad air is responsible for cutting the average lifespan of the global population by two years—a grim number that increases to up to six years in countries that struggle with severe air pollution, such as India.

3   The new guidelines recommend lower levels for six pollutants: ozone, sulphur dioxide, carbon monoxide, nitrogen dioxide and particulate matter (PM10 and PM2.5).

4   The new limit for nitrogen dioxide (NO2), among the most harmful particles, is now 75 percent lower than in 2005.

5   The PM2.5 guideline level has been halved, after evidence has emerged that these particles can enter not only deep into the lungs, but also into the bloodstream, causing mainly cardiovascular and respiratory problems and affecting other organs.

6   "WHO has adjusted almost all the air quality guideline levels downwards, warning that exceeding the new levels is associated with significant risks to health," said the WHO.

7   "Adhering to them could save millions of lives."

**How harmful is air pollution?**

8   At least 7 million people are killed by air pollution every year, according to data collected by the WHO.

9   Recent research estimated that between 2012 and 2018, burning fossil fuels caused 8.7

million premature deaths across the world—including among thousands of children under five years old.

**10** Of the total premature deaths related to air pollution:
- 27 percent are due to pneumonia
- 18 percent are from stroke
- 27 percent are from ischemic heart disease
- 20 percent are from chronic obstructive pulmonary disease
- 8 percent are from lung cancer.

**11** In children, complications from air pollution often cause reduced lung growth and function, respiratory infections and aggravated asthma.

**12** Evidence has also emerged since 2005 that air pollution can be linked to other diseases such as diabetes and neurodegenerative conditions.

**13** A study from 2019 linked environmental pollution to the increased risk of developing bipolar disorder and depression.

**14** The WHO said air pollution, together with the climate crisis, is now one of the biggest environmental threats to human health.

**15** The UN body stressed that even sticking to the new guidelines, the global population remains exposed to damaging particles that are always harmful, no matter in what quantity they are present in the air.

**16** But if countries will manage to stick to the guideline levels recommended by the WHO, the organization estimates that almost 80 percent of deaths related to air pollution could be avoided.

**17** It will be up to each nation to decide whether to comply with WHO's guidelines on air pollution.

**What are the main contributors to air pollution in our cities?**

**18** In 2019, more than 90 percent of the world's population lived in areas where concentrations exceeded the 2005 guidelines for long-term PM2.5 exposure.

19    "Almost everyone around the world is exposed to unhealthy levels of air pollution," said WHO Director-General Tedros Adhanom Ghebreyesus.

20    "Inhaling dirty air increases the risk of respiratory diseases like pneumonia, asthma... and increases the risk of severe COVID-19."

21    Air pollution—a mix of particles and gases that are harmful to human health either indoors or outdoors—has several contributing factors that could partly be attributed to modern life.

22    There's a link between the climate crisis and air pollution: most of the pollutants in the atmosphere come from burning fossil fuels for energy and transportation.

23    Historically, the main pollutant in both developed and industrializing countries has been high levels of smoke and sulphur dioxide coming from domestic and industrial use of coal, oil and gasoline.

24    These days, the major contributing factor to air pollution is traffic emissions, especially in urban areas.

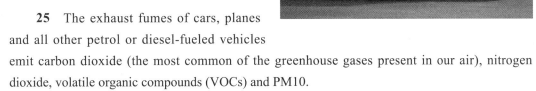

25    The exhaust fumes of cars, planes and all other petrol or diesel-fueled vehicles emit carbon dioxide (the most common of the greenhouse gases present in our air), nitrogen dioxide, volatile organic compounds (VOCs) and PM10.

26    Power plants and refineries emit more sulphur dioxide nowadays than volcanoes, which historically were the main emitters of this gas.

27    Other pollutants come from landfills and livestock (methane), using chemical and synthetic cleaning products at home (which emit VOCs) and wildfires.

28    Wildfires, which have devastated California in the U.S., Italy, France, Greece, Spain, Morocco and other places across the world, have not only destroyed hundreds of hectares of land, but they've also released PM2.5 in the air which, colliding with other harmful particles, contributes to smog.

**TEXT B**

## Eco-Friendly Living Becomes the Prevailing Ethos

China's State Council Information Office

1    Green development requires everyone's efforts, and each of us can promote and practice

green living. China actively promotes the values and ideas of eco-environmental conservation, raises public awareness to conserve resources and protect the eco-environment, and advocates the practice of a simpler, greener, and low-carbon lifestyle, creating a conducive social atmosphere for jointly promoting green development.

**Continuing progress towards raising conservation awareness**

2   China places particular emphasis on cultivating its citizens' conservation awareness. It organizes systematic publicity and other awareness-raising activities in this regard, and advocates a social environment and lifestyle of diligence and frugality. Publicity activities themed on National Energy-Saving Publicity Week, China's Water Week, National Urban Water-Saving Week, National Low-Carbon Day, National Tree-Planting Day, World Environment Day, the International Day for Biological Diversity, and Earth Day, are organized on a regular basis to encourage and persuade the whole of society to engage in green development activities. The idea of eco-friendly living has become widely accepted in families, communities, factories, and rural areas.
Material on green development has been incorporated into China's national education system through compiling textbooks on eco-environmental conservation and carrying out education in primary and secondary schools on the condition of national resources including forests, grasslands, rivers and lakes, land, water and grain. Respect for and love of nature have been advocated. Environmental Code of Conduct for Citizens (for Trial Implementation) was published to guide the public to follow a green lifestyle. As a result, a culture of ecological and environmental protection has joined the mainstream and been cherished by all.

**Widespread initiatives to promote eco-friendly lifestyles**

3   China has launched initiatives to promote the building of resource-conserving Party and government offices, and develop eco-friendly families, schools, communities, transport services, shopping malls, and buildings, popularizing eco-friendly habits in all areas including clothing, food, housing, transport, and tourism. To date, 70 percent of Party and government offices at and above county level are now committed to resource
conservation, almost 100 colleges and universities have realized smart monitoring of water and electricity consumption, 109 cities have participated in green transport and commutes initiatives. Household waste sorting has been widely promoted in cities at or above prefecture level. Much

progress is being made as residents gradually adopt the habit of sorting their waste. The Law of the People's Republic of China on Food Waste has been enacted, and initiatives launched to promote food saving and curb food waste including a "clean plate" campaign on a large scale, which have yielded remarkable results as more people are saving food.

**Growing market of green products**

4   China has actively promoted energy-saving and low-carbon products such as new-energy vehicles and energy-efficient household appliances. It has provided tax reductions or exemptions and government subsidies for new-energy vehicles and continued to improve charging infrastructure. As a result, the sales of new-energy vehicles have rapidly risen from 13,000 in 2012 to 3.52 million in 2021. For the seven years since 2015, China has ranked first in the world in the production and sales of new-energy vehicles. In addition, China has steadily improved the certification and promotion system for green products and the green government procurement system, implemented an energy efficiency and water efficiency labeling system to encourage the consumption of green products. It has promoted the construction of green infrastructure in the circulation sector such as green shopping malls, and supported new business models such as the sharing economy and second-hand transactions. There is a richer variety of green products and a growing number of people who spend on green products.

## TEXT C

微课

### World Must Better Protect Biodiversity

Lu Lunyan

1   Governments are meeting at the 15th Conference of the Parties to the Convention on Biological Diversity, in Montreal, Canada, from Dec 7-19 to agree on new global goals and a 2030 action plan for nature. COP 15 must set the world on a new course to address alarming biodiversity loss, which is also vital to combating the climate emergency, achieving food and water security, reducing our vulnerability to future pandemics, and achieving the Sustainable Development Goals.

2   The World Wide Fund for Nature is urging world leaders to secure an ambitious global agreement to save our life support systems at COP 15. Nature is declining at rates unprecedented in history, with 1 million species threatened with extinction. People are becoming increasingly aware of this crisis. New WWF research published this year shows the amount of people concerned about rapid nature loss in the world's top global biodiversity hotspots has risen to nearly 60 percent—reflecting a 10 percentage point increase, since 2018. In addition, nature and climate change were seen as the most important policy areas for people (81 percent), across the thousands surveyed.

3   As humans continue to exploit and destroy nature on an unprecedented scale, we undermine the living systems that our own well-being, security and prosperity depend upon. The post-2020 Global Biodiversity Framework needs to be at least as comprehensive, science-based and ambitious as the Paris Agreement on Climate Change. It must deliver immediate action on the ground to reverse nature loss; however, the current draft framework does not go far enough to address the biodiversity crisis. WWF has identified a number of areas in which the new Global Biodiversity Framework must galvanize transformative action.

4   All countries must increase ambition and action to deliver a comprehensive and science-based framework that can halt and reverse biodiversity loss by 2030 and achieve a nature-positive future, so that by the end of the decade we have more biodiversity than at its start. The GBF needs to provide a big advance on the previous 10-year strategic plan for the CBD—under which none of the 20 so-called Aichi targets  were fully met. It was the second consecutive decade that the world failed to meet biodiversity targets. During this time, biodiversity loss has continued, at an increasing rate. Governments must set new ambitious targets for 2030 and be held accountable for meeting them.

5   COP 15 must result in an agreement on a strong and ambitious GBF that allows parties to begin immediately implementing it by setting up national targets and updating National Biodiversity Strategies and Action Plans in line with the new global framework. This will require the adoption at COP 15 of a package that, at a minimum, includes, in addition to the post-2020 GBF document:

The monitoring framework including headline indicators;

The enhanced multidimensional approach to planning, monitoring, reporting and review;

The strategy for resource mobilization.

6   The COP decision which operationalizes the GBF should identify that the immediate and full implementation of the GBF, including through adequate national legislation, is required by parties in order to fulfill their obligations under the convention.

7   To address our escalating nature crisis and secure a sustainable future for current and future generations, the adopted framework must, at a minimum, contain the following key negotiated policy outcomes.

8   WWF will be pressing governments in Montreal to adopt a "Paris"-style agreement capable of driving immediate action to halt and reverse biodiversity loss by 2030 for a nature-positive world. This means having more biodiversity at the end of the decade, than we have now.

9   To date, more than 90 world leaders have endorsed the Leaders' Pledge for Nature, committing to reverse biodiversity loss by 2030.

10   WWF stresses the importance of countries agreeing to a goal of conserving at least 30 percent of the planet's land, inland waters and oceans by 2030.

11   At the same time, action is needed to ensure the remaining 70 percent of the planet is sustainably managed and restored—and this means addressing the drivers of biodiversity loss, with the same level of urgency. Science is clear that global production and consumption rates are completely unsustainable and are causing serious damage to the natural systems people rely on for their livelihoods and well-being. WWF believes a commitment to half the global footprint of production and consumption by 2030, while recognizing huge inequalities between and within countries, is desperately needed in the framework to ensure that key sectors, such as agriculture and food, fisheries, forestry, extractives and infrastructure, are transformed to help deliver a nature-positive world.

12   Despite a large and growing number of world leaders committing to secure an ambitious global biodiversity agreement, key issues remain unresolved, including how to mobilize the necessary finance. Currently, the biodiversity finance gap is estimated at $700 billion annually. WWF is encouraged by the Kunming Biodiversity Fund that China initiated. WWF is calling for countries to substantially increase finance, including international public finance with developing countries as the beneficiaries, and align public and private financial flows with nature-positive practices, including through the elimination or repurposing of harmful subsidies and other incentives.

**13** The talks are the finale to what has been an incredibly challenging four years of negotiations, with the pandemic delaying any agreement on the global biodiversity framework under the UN Convention of Biological Diversity until now.

**14** A strong implementation mechanism which requires countries to review progress against targets and increase action as required is an essential mechanism to ensure real action is delivered on the ground.

**15** This proposal is underlined by WWF's research, which found 56 percent of people surveyed believe government action to protect biodiversity is insufficient. The research, which surveyed more than 9,200 people across regions with staggering rates of biodiversity loss, found that people also perceived policy-related actions to be more impactful than individual consumer action.

## Writing

Choose one endangered animal or plant you're interested in and search for its information online. What's its importance to biodiversity? What measures have been taken to protect them?

## Unit Project

Work in groups and make a flow chart of your daily activities. Analyze the activities, and discuss what kinds of pollutants you may generate, the impacts on the environment and what could be changed to alleviate the negative effects. Present your report to the whole class.

# Unit 3

# Population, Urbanization and Ecology

Urbanization is a global trend. The UN predicts that two out of every three people will likely be living in cities or on the urban centers by 2050. The fast expansion of cities has imposed unprecedented challenges in sustainability and environmental protection. The challenges can be environmental degradation and urban poverty. The goal of sustainable cities and communities has been highlighted at UN Sustainable Development Goals (SDGs), ranking the 11th among the 17 goals. However, the World Bank warns that large cities in many developing countries now face excessively high population levels. Alleviating the urban population strain in developing countries requires equality in services and environmental protection. Policy makers, urban planners and service providers are expected to promote and contribute to sustainable urbanization with cross-cutting approach. How to make our cities more sustainable? What initiatives should our government and businesses put in place? Chinese cities are turning things around with many considerably lowering emissions in the past year, while also maintaining significant economic growth. This turnaround comes as industries and buildings have started using clean energy, transport and adopting smart technologies. Urban development led by the industrial and service sectors has also pulled more than 700 million people out of poverty in the last 30 years.

# Practical English for Ecological Environment

## Warm-up

Today more than half of all people in the world live in an urban area. By mid-century, this will increase to 70%. How have we reached such a high degree of urbanization and what does it mean for our future?

You are going to watch a video about urbanization and the future of cities. Watch it twice and complete the following two tasks. Compare your answers with your partner's.

**Task 1**  Watch the video clip and take notes according to the clues given.

(1) Birth of cities
(2) Main factors of the development of cities
(3) Main causes of the high density of ancient cities
(4) Modern cities
(5) Suggestions for future cities

**Task 2**  Watch the video clip again and fill in the blanks with the missing information. You can refer to the notes you have taken.

(1) About 10,000 years ago, our _____ began to learn the secrets of _____ and _____.

(2) It was only with _____ like _____ about 5000 years ago that people could _____ food, making permanent settlements possible.

(3) And as trade _____, so did technologies that _____ it.

(4) Modern cities didn't really get started until _____ when new technology deployed _____ allowed cities to _____ further.

(5) The world will need to seek ways to _____ for all people. Power will increasingly come from _____.

**Task 3**  Work in small groups. Discuss with your group members the following questions.

(1) Describe a city you have been to or lived in China.
(2) If you were to travel to a city in a foreign country, which one would you choose and why?

## Reading Comprehension

TEXT A

### ASEAN's Rapid Urbanization Faces Sustainability Woes

Alok Gupta

1  Countries in the Association of Southeast Asian Nations (ASEAN) will **witness** massive urbanization in the next three decades, making them the world's largest middle-income emerging markets after China and India, and leading to enormous resource scarcity, a new report **claims**.

2  The United Nations Environment Programme (UNEP) **estimates** that more than 205 million people in Indonesia, the Philippines, Vietnam, Thailand, Myanmar, Malaysia, Cambodia, Laos, Singapore, and Brunei <u>are expected to</u> move to cities by 2050, <u>according to</u> the report "ASEAN Region: A Resource Perspective."

3  <u>At present</u>, 320 million people live in the urban region of ASEAN countries, and by 2050 this population will **swell** to nearly 525 million. The **unprecedented** urban growth is expected to lead to a rapid rise of more than 200 smaller cities.

4  The region's urbanized population proportion will increase from 47 percent in 2014 to 65 percent in 2050, with five of the 10 ASEAN nations transitioning from a minority urban population to a majority urban population, the report predicted.

(Total projected urban population in ASEAN, India and China in 2050. ASEAN urbanization is projected to be one of the largest after China and India, particularly in the context of urban middle class expansion)

**Population growth likely to lead to an increase in pollution and resource crunch**

5   Researchers warn that rapid urbanization in the ASEAN region could lead to dangerous levels of air pollution, similar to those seen in several cities in India and China with average annual PM2.5 **concentrations exceeding** 100 micrograms.

6   Air pollution in the region will be further **exacerbated** by the 120 million who don't have access to electricity and the 280 million who lack clean cooking fuels. The addition of waste burning will only fuel this environmentally damaging trend.

7   With the rise of an urban population, the report predicts, the share of coal in the region's electricity supply mix will rise from 32 percent to a whopping 50 percent by 2040 to meet the increase in energy demand leading to a **spike** in emissions.

8   "Today, most ASEAN cities report PM2.5 levels above those that the World Health Organization's (WHO) safe limit of 10 micrograms," report maintains.

9   "There are already far too many people around the world who are already being poisoned by breathing dirty, dangerous air in the cities they live in, and it's alarming to see that this trend is set to worsen," said UN Environment chief Erik Solheim.

10   Apart from toxic air, urban land expansion is likely to reduce total crop production by two percent in Indonesia and 16 percent in Vietnam as agricultural land reclamation will make way for urban development. The Philippines will lose more than four percent of its biodiversity hotspots.

**UN   suggests ASEAN countries act before it's too late**

11   UNEP has urged policymakers in ASEAN countries to take immediate action in five areas based on the New Urban Agenda (NUA) and Sustainable Development Goals (SDG) for better urbanization.

12   Researchers recommend national and cross-ASEAN urbanization planning to balance economic growth across a range of city sizes while **preserving** high-value agricultural lands and ecosystem services.

**Examples of fast growing cities in the ASEAN region 2015-2030**

| Country | City | Absolute Percentage Growth 2015-2030 | Annual Percentage Growth Rate 2015-2030 |
|---|---|---|---|
| Laos | Vientiane | 78.8 | 5.3 |
| Indonesia | Batam | 78.8 | 5.3 |
| Thailand | Samut Prakan | 73.1 | 4.9 |
| Indonesia | Denpasar | 69.0 | 4.6 |
| Indonesia | Tasikmalaya | 65.8 | 4.3 |
| Vietnam | Can Tho | 61.9 | 4.1 |

**13** For better management of land, city planners are advised to promote compact, mixed-use, accessible, and **inclusive** urban form to reduce land expansion, **streamline** infrastructure provision, and promote diverse, **sustainable** mobility options.

**14** Planners have been suggested to consider resource-efficient and **resilient** buildings and electric-grid systems that link renewable energy in cities with the cross-ASEAN electric grid for efficient energy consumption.

**15** "We can and need to do far better. We can design better cities, where people can walk or cycle instead of having to use cars, where waste is recycled rather than burned or tossed into landfills, and where everyone can access clean fuels and energy," Solheim added.

**16** The report also mentions the Indian city of Ahmedabad as an example of a success story in urban and regional planning that has kept pace with rapid population growth.

**17** The city's municipal development authority has been **competent** in using land pooling to manage urban land expansion and reduce sprawl compared to other large cities in India, researchers maintained.

**18** UNEP **released** the report during the ongoing World Urban Forum in Kuala Lumpur, Malaysia.

**19** "The ASEAN region will continue to grow as an economic powerhouse, projected to become the fourth largest economy in the world by 2050," the report said.

### New Words

**woe** /wəʊ/ *n.*
the troubles and problems that sb has 麻烦；问题；困难
**witness** /ˈwɪtnəs/
*n.*
a person who sees sth happen and is able to describe it to other people 目击者；见证人
*v.*
to see sth happen (typically a crime or an accident) 当场看到，目击（尤指罪行或事故）
**claim** /kleɪm/
*n.*
a statement that sth is true although it has not been proved and other people may not agree with or believe it 声明；宣称；断言
*v.*
to say that sth is true although it has not been proved and other people may not believe it 宣称；声称；断言
**estimate** /ˈestɪmeɪt/ *v.*

to form an idea of the cost, size, value, etc. of sth, but without calculating it exactly 估价；估算

**swell** /swel/ *v.*

to increase or make sth increase in number or size （使）增加，增大，扩大

**unprecedented** /ʌnˈpresɪdentɪd/ *adj.*

that has never happened, been done or been known before 前所未有的；空前的；没有先例的

**concentration** /ˌkɒns(ə)nˈtreɪʃ(ə)n/ *n.*

the amount of a substance in a liquid or in another substance 浓度；含量

**exceed** /ɪkˈsiːd/ *v.*

to be greater than a particular number or amount 超过（数量）

**exacerbate** /ɪɡˈzæsəbeɪt/ *v.*

to make sth worse, especially a disease or problem 使恶化；使加剧；使加重

**spike** /spaɪk/ *n.*

a sudden large increase in sth 猛增；急升

**preserve** /prɪˈzɜːv/ *v.*

to keep a particular quality, feature, etc.; to make sure that sth is kept 保护；维护；保留

**inclusive** /ɪnˈkluːsɪv/ *adj.*

including a wide range of people, things, ideas, etc. 包容广阔的；范围广泛的

**streamline** /ˈstriːmlaɪn/ *v.*

to make a system, an organization, etc. work better, especially in a way that saves money 使（系统、机构等）效率更高；（尤指）使增产节约

**sustainable** /səˈsteɪnəb(ə)l/ *adj.*

that can continue or be continued for a long time 可持续的

**resilient** /rɪˈzɪliənt/ *adj.*

able to feel better quickly after sth unpleasant such as shock, injury, etc. 可迅速恢复的；有适应力的

**competent** /ˈkɒmpɪtənt/ *adj.*

having enough skill or knowledge to do sth well or to the necessary standard 足以胜任的；有能力的；称职的

**release** /rɪˈliːs/ *v.*

to make sth available to the public 公布；发布

## Phrases and Expressions

**be expected to do:** be believed to happen 预期；有望

**according to:** as stated or reported by sb/sth 据（……所说）；按（……所报道）

**at present:** at the present moment 目前；现在

**have access to:** to have the opportunity or right to use sth or to see sb/sth （使用或见到的）机会，权利

**be set to:** be ready to, be prepared to (do something); to be on the point of (doing something) 注定；准备做

**apart from:** in addition to; as well as 除了……外（还）；此外；加之

**be likely to:** is probable to or expected to 可能的；预料的；有希望的

**make way for:** to provide a space or an opportunity for something else 让路给；让步于

**take action:** to act in order to get a particular result 采取行动

**rather than:** rather than introduces the thing or situation that is not true or that you do not want 而非；而不是

**keep pace with:** to go or make progress at the same speed as (someone or something else) 并驾齐驱；保持同步

**projected to:** be planned to happen in the future 预计

### Exercises

**I Understanding the Text**

1. Recognize the sequence in which the text presents its most important information on challenges brought by urbanization. Fill in the form with the correct choice form the following list.

① UN's suggestion that ASEAN countries act before it's too late

② Population growth likely to lead to an increase in pollution and resource crunch

③ The prediction of the trend of massive urbanization in ASEAN

④ The trend of massive urbanization in ASEAN and its effects

⑤ The ASEAN region's growth as an economic powerhouse

| Parts | Paragraphs | Main ideas |
| --- | --- | --- |
| Part One | Para. 1 | |
| Part Two | Paras. 2-4 | |
| Part Three | Paras. 5-10 | |
| Part Four | Paras. 11-17 | |
| Part Five | Paras. 18-19 | |

2. Focus on the sentences that provide impressive data and fill in the blanks. Pay attention to the verbs, nouns, adjectives and adverbs commonly used when giving statistics.

① The UNEP _____ in Indonesia, the Philippines, Vietnam, Thailand, Myanmar, Malaysia, Cambodia, Laos, Singapore, and Brunei are expected to move to cities by 2050.

② _____ live in the urban region of ASEAN countries, and by 2050 this population will _____.

③ The unprecedented urban growth is expected to lead to _____.

④ The region's urbanized population proportion will _____, with _____ ASEAN nations transitioning from a minority urban population to a majority urban population, the report predicted.

⑤ Population growth is likely to _____ pollution and resource crunch.

⑥ Urban land expansion is likely to _____.

**II Focusing on Language in Context**

1. Key Words & Expressions

A. Fill in the blanks with the words given below. Change the form where necessary. Each word can be used only once.

**witness unprecedented exacerbate competent release sustainable spike concentration estimate preserve**

① The 1980s _____ an unprecedented increase in the scope of the electronic media.

② He was a loyal, distinguished and very _____ civil servant.

③ The creation of an efficient and _____ transport system is critical.

④ We will do everything to _____ peace.

⑤ The symptoms may be _____ by certain drugs.

⑥ Police have _____ no further details about the accident.

⑦ We are seeing unemployment on an _____ scale.

⑧ The recent _____ in food prices around the world that you've all heard of is because of rising energy costs.

⑨ Tiredness affects your powers of _____.

⑩ It is _____ that total investment in the country will continue to increase this year.

B. Fill in the blanks with the phrases given below. Change the form where necessary.

**according to; apart from; make way for; rather than; keep pace with; be projected to; at present; take action; have access to; be set to**

① You've been absent six times _____ our records.

② The report was a device used to hide _____ reveal problems.

③ She found it hard to _____ him as he strode off.

④ 13% of Americans are over 65 and this number _____ reach 22% by the year 2030.

⑤ There are also situations in which you might need to jump a red light, such as to _____ _____ a fire engine.

⑥ We need more time to see how things develop before we _____.

⑦ As things stand_____, he seems certain to win.

⑧ His fridge was bare _____ three very withered tomatoes.

⑨ The drug _____ become the treatment of choice for asthma worldwide.

⑩ Millions of people cannot read these words because they don't _____ a computer.

2. Usage

A. Complete the following sentences in the text by using "with" phrases.

① _____, the report predicts, the share of coal in the region's electricity supply mix will rise from 32 percent to a whopping 50 percent by 2040 to meet the increase in energy demand leading to a spike in emissions.

② The region's urbanized population proportion will increase from 47 percent in 2014 to 65 percent in 2050, _____ five of the 10 ASEAN nations _____, the report predicted.

③ Researchers warn that rapid urbanization in the ASEAN region could lead to dangerous levels of air pollution, similar to those seen in several cities in India and China average annual PM2.5 concentrations_____.

B. Filling the blanks with "rather than" based on the Chinese.

① "We can design better cities, where people can walk or cycle instead of having to use cars, where waste is recycled _____,（而不是烧掉或扔进垃圾填埋场）and where everyone can access clean fuels and energy," Solheim added.

② It now seems a probability _____.（而不是仅有可能）

③ The test was to assess aptitude _____.（而不是学业成绩）

④ It's far better to teach children how to do these things themselves _____.（比一直为他们做这些事情）

**III Translation**

Translate the following passage into English.

当前，城镇化（urbanization）的全球趋势以及世界人口稳步增长的趋势已经持续了很长时间。对于发达国家来说，没有迹象表明城镇化会导致人口增长（population growth），但在发展中国家，城镇化和人口增长则紧密相关。城镇化对中国有某种积极的影响，随着越来越多的人集中在城市寻找工作或商业机会，工业也随着大量劳动力而繁荣起来。

**IV Pair Work**

Talk about the causes of urbanization and the benefits, problems and challenges brought by it.

**TEXT B**

## A Model for Urbanization: China's City Clusters

Han Jie

1   Top of the **agenda** for this year's Asian-Pacific Economic Cooperation (APEC) is <u>the process of</u> urbanization and its role in developing the regional economy.

2   The **summit**, which started on Monday in Vietnam, has implemented a policy of "Five Nos" and "Three Haves", with the "haves" <u>referring to</u> the goal of people having houses, jobs and urban **civilized** lifestyle.

3   <u>With the population shift from rural to urban areas, urbanization can seem a necessary path toward modernization as well as a key engine for sustainable economic growth.</u>

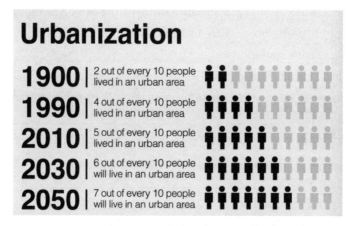

4   Working to connect the region through improving physical **infrastructure** linkages, mobility and institutional ties, APEC is also working to ensure all members can <u>participate in</u> the growing economy by learning from other regions.

**Chinese model of urbanization**

5   In China, more than 80 million people—about 57 percent of the Chinese population are currently living in cities. As the world's second-largest economy, China continued its firm growth in the first three quarters of this year, with gross domestic product expanding 6.9 percent year-on-year to 59.3288 trillion yuan (about 8.9614 trillion US dollars), the National Bureau of Statistics report shows.

6   <u>Although China's extraordinary economic **boom** has gone hand-in-hand with urbanization, it has led to a sharp increase in consumer and investment demand, and huge needs in infrastructure, public service and housing.</u>

7   According to a recent report published by the Ministry of Human Resources and Social

Security (MHRSS), the registered unemployment rate in Chinese cities stood at 3.95 percent at the end of the third quarter, the lowest level since 2008.

8   China's urbanization program is an important part of structural reform, as the nation transitions to a more productive, service-based economy. General Secretary Xi Jinping stressed the importance of developing city clusters—coordinated regional development and urbanization—at the 19th National Congress of the Communist Party of China this year, mentioning that China's urbanization is at a crucial **juncture**, and it now needs to focus on these clusters.

9   In order to **combat** the **sluggish** global economic recovery and rising challenges and risks, urbanization in the Asia-Pacific economies needs to not only **drive** market demand but can also help achieve wider coverage of public services, enhanced food security, better environmental protection, narrower urban-rural gaps and inclusive economic growth.

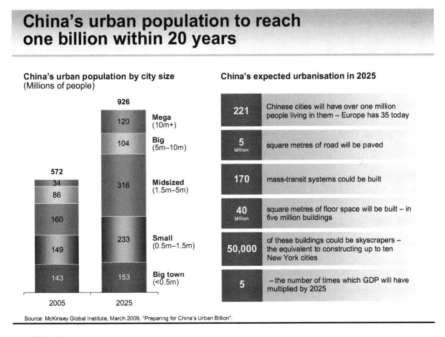

**Xiongan New Area**

10   Like China's urbanization plan, it has been upgraded from individual cities and provinces to **world-class** city clusters, according to the 19th CPC National Congress.

11   In an ideal city cluster, big cities are connected with each other by inter-city railways, expressways and other transport means, which also connect them with nearby smaller towns. In the Xiongan New Area, the coordinated development of Beijing, Tianjin and Hebei Province is a national strategy that aims to create a world-class city cluster in northern China.

12   Compared with the Yangtze River and the Pearl River Deltas, where the coordination and integration levels are much higher, the Xiongan New Area will integrate the economies of

the Beijing-Tianjin-Heibei region into the mega-region Jing-Jin-Ji. The region is designed to hold 110 million people and **merge** outer parts of the three areas in order to break down "fortress economies" in the region.

**13**   He Lifeng, director of China's national development and reform commission, said in March that Jing-Jin-Ji would help distribute wealth that had been sucked into Beijing to the areas surrounding the capital—many of which are still **impoverished** villages.

**Urbanization and Chinese supply-side reform**

**14**   With corporate universities, institutions and residents from the capital, helping to **alleviate** pressure on housing and public services, the Jing-Jin-Ji area hopes to strengthen the coordination of local governments and provide integrated public services to locals.

**15**   The integrated development of urban and rural areas is expected to generate investment and consumption of nearly 100 trillion yuan, which "will be the most remarkable bonus for China's development in the medium to long run", said Chi Fulin, head of the China Institute for Reform and Development.

**16**   While gains from structural reforms will take time, they will have a positive impact on China's economic growth in the medium term, said Changyong Rhee, director of the Asia Pacific Department at the IMF, adding that China's growth has also provided **ample** opportunities for Asia to maintain its growth over the last 10 years.

**17**   From field to the office, China's high efficiency in decision-making and **execution** means it can **mobilize** the necessary resources in a short time to foster world-class urban clusters.

### New Words

**cluster** /ˈklʌstə(r)/ *n.*
a group of people, animals or things close together （人或动物的）群，团，组
**agenda** /əˈdʒendə/ *n.*
a list of items to be discussed at a meeting （会议的）议程表，议事日程
**summit** /ˈsʌmɪt/ *n.*
an official meeting or series of meetings between the leaders of two or more governments at which they discuss important matters （政府间的）首脑会议；峰会
**civilize** /ˈsɪvəlaɪz/ *v.*
to educate and improve a person or a society; to make sb's behaviour or manners better  教化；

开化；使文明；使有教养

**infrastructure** /ˈɪnfrəstrʌktʃə(r)/ *n.*

the basic systems and services that are necessary for a country or an organization to run smoothly, for example buildings, transport and water and power supplies （国家或机构的）基础设施，基础建设

**boom** /buːm/ *n.*

a sudden increase in trade and economic activity; a period of wealth and success （贸易和经济活动的）激增，繁荣

**juncture** /ˈdʒʌŋktʃə(r)/ *n.*

a particular point or stage in an activity or a series of events 特定时刻；关头

**combat** /ˈkɒmbæt/ *v.*

to stop sth unpleasant or harmful from happening or from getting worse 防止；减轻

**sluggish** /ˈslʌɡɪʃ/ *adj.*

moving, reacting or working more slowly than normal and in a way that seems lazy 缓慢的；迟缓的；懒洋洋的

**drive** /draɪv/ *v.*

to influence sth or cause it to make progress 激励；促进；推进

**world-class** /ˌwɜːld ˈklɑːs/ *adj.*

as good as the best in the world 世界级的；世界上一流的

**merge** /mɜːdʒ/ *v.*

to combine or make two or more things combine to form a single thing （使）合并，结合，并入

**impoverish** /ɪmˈpɒvərɪʃ/ *v.*

to make sb poor 使贫穷

**alleviate** /əˈliːvieɪt/ *v.*

to make sth less severe 减轻；缓和；缓解

**ample** /ˈæmpl/ *adj.*

enough or more than enough 足够的；丰裕的

**execution** /ˌeksɪˈkjuːʃn/ *n.*

the act of doing a piece of work, performing a duty, or putting a plan into action 实行；执行；实施

**mobilize** /ˈməʊbəlaɪz/ *v.*

to find and start to use sth that is needed for a particular purpose 调动；调用

## Phrases and Expressions

**the process of:** a series of things that are done in order to achieve a particular result （为达到某一目标的）过程；进程

**refer to:** to mention or speak about sb/sth 提到；谈及；说起

**participate in:** to take part in or become involved in an activity 参加；参与

**stand at:** to be at a particular level, amount, height, etc. 达特定水平（或数量、高度等）

**integrate...into...:** to combine two or more things so that they work together; to combine with sth else in this way （使）合并，成为一体

**break down:** to destroy sth 搞垮

**take time:** to need or require a particular amount of time 需要时间

### Exercises

**I Comprehension Check**

① What is the meaning of "Three Haves"?

② How many people in China are living in cities?

③ What is an important part of China's structural reform?

④ What are the benefits of urbanization for the Asia-Pacific economies?

⑤ What will be the most remarkable bonus for China's development?

**II Translation**

Translate into Chinese the underlined sentences in the article.

① With the population shift from rural to urban areas, urbanization can seem a necessary path toward modernization as well as a key engine for sustainable economic growth.

② Although China's extraordinary economic boom has gone hand-in-hand with urbanization, it has led to a sharp increase in consumer and investment demand, and huge needs in infrastructure, public service and housing.

③ In an ideal city cluster, big cities are connected with each other by inter-city railways, expressways and other transport means.

④ While gains from structural reforms will take time, they will have a positive impact on China's economic growth in the medium term.

⑤ China's high efficiency in decision-making and execution means it can mobilize the necessary resources in a short time to foster world-class urban clusters.

**III Group Work**

China has made great progress in both urban construction and rural revitalization. Illustrate either progress with examples in your hometown.

**TEXT C**

## Energy Transformation Key to Clean Heating

Han Wenke

1   Replacing fossil fuels with clean energy has become an **irreversible** trend in China, and with the introduction of a series of policies to **peak** its carbon emissions before 2030 and achieve carbon neutrality before 2060, its low-carbon energy transformation is constantly accelerating.

2   That shows that energy transformation, which is part of efforts to actively respond to climate change, is gaining **momentum** in China.

3   Because of the amount of land, labor, capital, and entrepreneurship that it possesses and can exploit for manufacturing, its status as a developing country, and its stage of development, China remains the world's largest producer and consumer of fossil fuels. However, as climate change and energy issues are becoming more of a challenge, China is upholding energy transformation with a revolutionary **zeal**.

4   After President Xi Jinping proposed vigorously promoting an energy production and consumption revolution in 2014, China has accelerated reforms on the energy supply side and

consumption side, upgraded its energy technology and energy system, and accelerated the green and low-carbon transformation of China's industrial sectors and the entire economy.

5   For example, from 2016 to 2020, China has resolved 170 million metric tons of excess steel production capacity, removed 1 billion tons of excess coal production capacity, shut down 300 million tons of cement production overcapacity, and reduced flat glass production capacity by 150 million weight boxes.

6   The proportion of coal consumption in China's total primary energy consumption has dropped from 67 percent in 2005 to 56.8 percent in 2020, and the proportion of non-fossil fuels in its primary energy consumption has risen to 15.9 percent. China's investment, installed capacity, electric energy production and number of technical patents in renewable energy sources have ranked first in the world for many years.

7   Last year, China introduced the 1+N policy system— "1" being a master guideline issued by the central authorities, and "N" standing for specific action plans or policies for different industrial sectors —to achieve carbon peaking before 2030 and carbon neutrality before 2060. To realize these goals, the country has released action plans for an energy green transformation initiative, industrial carbon peak  initiative, green transportation low-carbon action, circular economy carbon reduction actions and other key areas and implementation plans to peak emissions in various industries, as well as a series of supporting measures.

8   These policies and measures have **expedited** China's development of clean energy, mainly solar and wind energy, and accelerated the pace of green low-carbon transformation, injecting new **impetus** into China's energy transformation. Meanwhile, it also expands space for the clean development of electric vehicles, hydrogen energy, energy storage and various distributed energy sources, including winter heating, to ensure people's livelihoods.

9   In recent years, some inefficient heating methods, such as scattered coal heating in rural areas, are gradually being replaced with electric heating and more efficient centralized heating; electric heating equipment for decentralized heating in some areas are being replaced with more efficient household heat pump equipment.

10   At the same time, some qualified urban and rural areas have implemented energy-saving renovation of old houses to improve the **insulation**, build energy efficiency, and correspondingly reduce the heat energy demand for winter heating.

**11**  In some remote areas, replacing coal-fired heating with electric heating has also effectively improved the local ecology. For example, Sanjiangyuan, which means the source of three rivers, in Northwest China's Qinghai Province is the highest and biggest plateau wetland in the world, and it is also an important water conservation site. For a long time, locals there have <u>relied on</u> burning coal and cow dung for heating, which, apart from low energy efficiency and heating quality, was also causing pollution. So the Sanjiangyuan Clean Heating Construction and Renovation Project was launched in May 2020. Under a pilot project, implemented gradually, traditional coal-fired soil boilers were gradually transformed into efficient and clean electric boilers. Qinghai Province has also implemented clean heating in public places such as schools and rural pastoral areas.

**12**  As China vigorously promotes its energy transformation to achieve its dual carbon goals, the building area of such clean heating systems will expand. Among the sources for regional central heating, the heat pump technology that is largely applied in China in recent years will play a dominant role. Heat pumps will replace fossil fuel-based boilers and, combined with electricity, make the central heating system green and low-carbon.

**13**  In short, with energy transformation, the use of energy will be more efficient, clean, affordable and convenient

### New Words

**irreversible** /ˌɪrɪˈvɜːsəbl/ *adj.*
that cannot be changed back to what it was before 无法复原（或挽回）的；不能倒转的
**peak** /piːk/ *v.*
to reach the highest point or value 达到高峰；达到最高值
**momentum** /məˈmentəm/ *n.*
the ability to keep increasing or developing 推进力；动力；势头
**zeal** /ziːl/ *n.*
great energy or enthusiasm connected with sth that you feel strongly about 热情；激情
**expedite** /ˈekspədaɪt/ *v.*
to make a process happen more quickly 加快；加速
**impetus** /ˈɪmpɪtəs/ *n.*
something that encourages a process or activity to develop more quickly 动力；促进；刺激
**insulation** /ˌɪnsjuˈleɪʃn/ *n.*
the act of protecting sth with a material that prevents heat, sound, electricity, etc. from passing through; the materials used for this 隔热；隔音；绝缘；隔热（或隔音、绝缘）材料

## Phrases and Expressions

**a series of:** several events or things of a similar kind that happen one after the other 一系列；连续；接连

**respond to:** to do sth as a reaction to sth that sb has said or done 做出反应；响应

**shut down:** if a company, factory, large machine etc shuts down or is shut down, it stops operating, either permanently or for a short time 关闭；停业

**inject into:** to add a particular quality to sth （给……）添加/增加（某品质）

**rely on:** to need or depend on sb/sth 依赖；依靠

## Exercises

**Comprehension Check**

① What has been an irreversible trend in China?

② What was the proportion of China's coal consumption in 2005?

③ What is 1+N policy?

④ What are the substitutes of some inefficient heating methods?

⑤ What will play a dominant role among the sources for regional central heating?

# Supplementary Reading

### TEXT A

### Hangzhou Embraces Fast Development of Digital Life

Hu Chao

1    China's digital economy is booming. And life has become much more convenient. The digital development has been driven in part by e-commerce giant Alibaba, which is headquartered in Hangzhou, the capital city of east China's Zhejiang Province.

2    Sitting near the beautiful West Lake, Hangzhou is one of China's leading e-commerce centers—one that has played a seminal role in digital development. Digital life in this lakeside city has been flourishing in recent years.

3    Instead of cash, cards and your wallet, you often only need your mobile phone and your "face" wherever you go and whatever you do in Hangzhou.

4   Many hotels in Hangzhou now have facial recognition systems. A quick scan of your face is all it takes for the machine to confirm your ID. So you can easily check in, even if you don't have your ID card on hand.

5   And the city has the country's first restaurant with a facial recognition system. Your face will be the one paying for the meal. Once you have chosen your meal on the big touch screen, all you need to do is type in your phone number, and confirm payment by scanning your face.

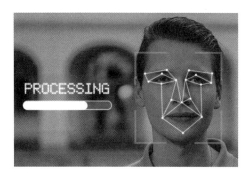

6   Many hospitals in Hangzhou also have become quite "smart". Long lines, stressful waits, huge crowds—the worst things about seeing a doctor—are things of the past in Hangzhou. In a local hospital, registration and payment can all be done through mobile apps, even before you arrive. It not only saves time, but also makes the hospital less crowded.

7   And the local central office of government affairs also sees fewer visitors now. Officials say the numbers have dropped by half because over 90 percent of administrative affairs now can be done either online or via these intelligent machines.

8   In May, Hangzhou set up the country's first Internet court. You don't even need to show up at court if you ever need to face the music in a lawsuit. This Internet court allows you to just log on and fight it out online. Over 4,000 cases have been handled so far.

9   Digital technology also helps to ease traffic congestion in Hangzhou, thanks to a "big brain of big data" in the local public security system and traffic police station. Hangzhou is the only city in China that has this "big brain".

10   Local authorities say that in the future this "big brain" will increasingly help to manage more aspects of the city, making life more convenient, efficient, and enjoyable.

**TEXT B**

微课

## Opinion: Structural Problems Challenge China's Rapid Urbanization

1   China's urban population has grown to more than 813 million, reaching nearly 60 percent of the country. The 2014-2020 New Urbanization Plan focuses on expanding cities, modernizing the countryside and building urban centers in rural regions.

2    The rapid urbanization achievement has been widely recognized. Uwe Brandes, the director of the Master's program in Urban and Regional Planning at Georgetown University, expressed his deep impression from a recent visit to China.

3    "I was surprised by the outstanding transportation infrastructure that has been built across the country," he said in a recent panel. "Major trends of innovation are happening in cities. The payment structure, for instance, is driving the city towards digitalization."

4    Although humans have been building cities for around 5,000 years, the cities in China are unique in terms of their historic value.

5    "The contemporary cities in China have been around for a long time," said Uwe. "Wuhan, for example, just celebrated its 3,000 years' history as a trading city midway on Yangtze River. Now the Belt and Road Initiative opens up new opportunities for it."

6    Unlike the common thought that the urbanization process is entirely a top-down proposal raised by the central government, Chinese scholars pointed out that the supportive voice from Chinese people should not be underestimated.

7    "It meets the needs of people who want to come to the city for higher income, better public services and better amenities in urban lives," said Yan Song, an urban planning and transportation consultant in China and the director of the Chinese Cities Program at the University of North Carolina.

8    "The mega-transportation project, urbanized centers, college towns and public parks are all offering desired job opportunities."

9    But such rapid urbanization is not without its challenges. According to Lu Zhe, a research fellow at the International Monetary Institute, the urban retention rate of the migrant workers is only somewhere around 18 percent, meaning that land growth has outpaced the urbanization of people.

TEXT C

### China Helps Increase Global Food Security

Zhao Yimeng

**Country's experience in reducing waste, increasing output inspiring to world**

1    China's accumulated experience in reducing food loss and waste is being shared with the

world and can help developing countries improve their ability to reduce food loss and ensure food security, experts and officials said.

2  About 14 percent of the world's food is lost during the processes from production to retail, and reducing that loss by 1 percentage point would be equal to a 28 million metric ton increase in grain output, which could feed 70 million people for a year, according to the State of Food and Agriculture 2019 report released by the United Nations Food and Agriculture Organization.

3  According to the FAO, over 155 million people experienced acute food insecurity in 2020, the highest number in the past five years, because of conflict, extreme weather events and economic shocks related to COVID-19.

4  Wu Laping, a professor at China Agricultural University's College of Economics and Management, said food loss and waste is largely affected by economic development, with developing countries losing food mainly before it is retailed and developed countries after retail.

5  For instance, food loss and waste mostly occurs in harvesting and storage in Africa and South Asia. Up to 30 percent of food loss was found during farmers' storage in some areas in Africa, and 21 percent of food in the United States was wasted during daily consumption, Wu told Economic Daily.

6  Compared with the world average, the food loss and waste situation in China is comparatively better. China loses 35 million tons of grain before retail each year, an amount roughly equal to the overall grain output of Sichuan Province last year—35.8 million tons.

7  China's experience shows that increasing grain output while reducing food loss and waste is a significant way to ensure national food security. In November, the country issued an action plan on saving food in multiple procedures to add "invisible" high-quality farmland.

8  Meanwhile, the reduction in loss and waste saves land, water, fertilizer and pesticides, and thus protects the environment and reduces carbon dioxide emissions, helping sustainable development.

9  "Because different countries face food loss and waste in different procedures, each country should find measures based on its situation," Wu said.

10  Lesser developed economies should focus on saving food during harvest and storage, which can be achieved by promoting varieties that can resist unfavorable conditions, such as

weather and pests, and by improving modern agricultural machinery, high-quality storage and transportation equipment, he added.

11　China has made progress in saving food by promoting laws against food waste and action plans on saving food and reducing loss in the whole process from harvest to retail, Zhou Guanhua, of the National Food and Strategic Reserves Administration, said last month.

12　"We have built 5,400 postproduction service centers, covering 1,000 major grain-producing counties in the country, to provide services including cleaning, drying, storing, processing and retailing," Zhou said, adding that the daily drying capacity can reach 1.1 million tons.

13　The centers can greatly reduce food loss that occurs through poor storage facilities and limited drying ability after harvest, as some farmers in Northeast China still dry grain on the ground or pile it in their yards.

14　Innovation, technologies and infrastructure are critical to increasing the efficiency of food systems and to reducing food loss and waste.

15　Zhou said new technologies to better store grain have been applied in warehouses. "Large grain depots in the country can reduce the rate of grain loss during storage to less than 1 percent."

16　The special vehicles for bulk grain, devices for unloading grain trucks, and automatic scales have proved effective, he said.

**"Healthy eating"**

17　People's growing awareness of "healthy eating" instead of seeking out refined staple foods has gradually prompted processing companies to transition from overprocessing to moderate processing.

18　"China is developing key technologies and equipment for moderate processing to help solve the problem of nutrition loss and impaired protein function due to overprocessing."

19　In September, China hosted the International Conference on Food Loss and Waste in Shandong Province. The recent food action plan proposes to make the conference a regular event.

20　Ma Youxiang, vice-minister of agriculture and rural affairs, said China would take the conference as an opportunity to promote establishing an international cooperation mechanism for reducing food loss and working together to boost world food security.

**21** As the largest developing country, China has actively participated in global food and agriculture management, launching cooperative procedures in saving food with other countries.

**22** Under the projects of the FAO, the World Food Programme and other international organizations, China shared information, techniques and talent with other countries, especially developing ones, to improve their ability to reduce food loss, Ma said.

**23** Sui Pengfei, director of the ministry's department of international cooperation, said the country will strengthen its collaborative research and development of techniques and equipment applied to key procedures for saving food.

## Writing

Focus on one of the traditional or renewable energy sources. What are the pros and cons of the energy? Illustrate your ideas with specific examples that make use of the energy.

## Unit Project

Work in groups. Take a look at the following table demonstrating China's ambitious plan to change its energy structure for carbon-neutrality purpose. Each group focuses on one energy source, describes the change, lists the pros and cons, and discusses the effective measures China has taken and will take in future to reduce or increase the source of energy. Present your report to the whole class.

| Energy Source | 2025 | 2060 | % Change |
| --- | --- | --- | --- |
| Coal | 52% | 3% | − 94% |
| Oil | 18% | 8% | − 56% |
| Natural Gas | 10% | 3% | − 70% |
| Wind | 4% | 24% | + 500% |
| Nuclear | 3% | 19% | + 533% |
| Biomass | 2% | 5% | + 150% |
| Solar | 3% | 23% | + 667% |
| Hydro | 8% | 15% | + 88% |

# Unit 4

# Industry and Environment

Industry is a seedbed for entrepreneurship, business investment, technological progress, the upgrading of skills, and the creation of decent jobs. One of the prerequisites for industry to flourish in a sustainable manner is the availability of an assured supply of affordable and clean energy, together with improved resource efficiency. Pollution, climate change, habitat destruction and over-exploitation of natural resources such as fresh water and fisheries are doing great harm to human health, well-being and livelihoods, especially among poorer regions, and is undermining the prospects for a long-term resilient and robust economy. The risks of climate change are well documented and its impacts are already affecting people and ecosystems. Meeting the climate challenge requires industries and institutions—both public and private—to be able to assess and understand climate change, design and implement adequate policies and to work towards resource efficient societies and low emission growth. "Decoupling" natural resource use and environmental impacts from economic growth is a key requirement for overcoming the pressing challenge of growing resource consumption levels.

*Practical English for Ecological Environment*

## Warm-up

The 17 Sustainable Development Goals (SDGs), agreed globally by all 193 UN member states, represent an affirmation of European values. The SDGs call on all nations to combine economic prosperity, social inclusion, and environmental sustainability. The SDGs are intimately linked with the Paris Climate Agreement (which is incorporated in SDG 13). The SDGs and the Paris Climate Agreement should be viewed as a package, with the SDGs oriented towards 2030 and the Paris Agreement oriented towards climate-neutrality by 2050, with major progress by 2030.

**Task 1** Watch the video clip and take notes according to the clues given.

(1) Birth of the idea of managing natural resources
(2) A picture of the Earth
(3) The goal of SDGs
(4) The significance of dealing with climate change
(5) Two important points in Brundtland Report

**Task 2** Watch the video clip again and answer the following questions. You can refer to the notes you have taken.

(1) Why did early ancestors realize the importance of managing natural resources?
(2) What are the two salient features of Earth shown by the photo?
(3) What are the two findings shown by the picture?
(4) What makes SDGs special in both its goal and process of achieving the goal?
(5) Why is Goal 13 climate change the most urgent one?

**Task 3** Work in small groups and take a look at the following picture. Discuss with your group members the following topics.

(1) Choose one that you are most concerned with and illustrate it with specific examples.

(2) Explain the importance of the one you've chosen and the relationship with other goals that interest you.

## Reading Comprehension

### TEXT A

**COP27: Why Global Action Is Needed to Decarbonize Industries Everywhere**

Rana Ghoneim

1  Ahead of this year's COP27 in Egypt, industry and government representatives from 15 developing countries across Asia, Latin America and Africa met in a series of **consultations** about the challenges and opportunities they face in decarbonizing some of their most energy **intensive** industries like steel, cement and **concrete**.

2  A report from these consultations—which were organized by the UN Industrial Development Organization (UNIDO), where I work—will be released during COP27's Decarbonization Day (11 November) and should be widely-read by decision-makers across energy, environment and industrial sectors.

3  During these meetings, it was evident that the pace of progress so far is too slow and that puts us at real risk of not meeting global climate **commitments**. It simply won't be **sufficient** for industrialized countries to lower emissions within their **boundaries** and **enforce** restrictions for products entering their markets. This must happen everywhere.

4  Global action and new forms of inter-sectoral cooperation are urgently needed to address **critical** questions including: what are the opportunities for emissions reductions, and what is needed to **deliver** these reductions in the fastest and most economical way?

5  How do we speed up the development and implementation of new carbon-cutting technologies—and **ensure** that they are widely accessible and affordable, including to small and medium sized enterprises?

**6** Currently, many developing country governments do not have **reliable** and up-to-date data on the emissions of their different industries and how they compare internationally. Relatively little has been established so far in the way of **infrastructure** to **facilitate** the widespread introduction of new and emerging technologies for industrial decarbonization.

**7** Access to and know-how about low-carbon technologies is largely concentrated within industrialized countries and large multinational companies.

**8** This must change. For industrial decarbonization efforts to succeed, we need to see significantly increased investments in research and development into new technologies—but we also need to scale up the deployment of technologies that exist but are not yet widely available, including those for carbon capture, utilization and storage (CCUS).

**9** We also need to much more widely **implement** strategies and technologies that are already available and affordable—including on energy efficiency, which lowers the demand for energy including from renewable sources.

**10** This likely requires new funding for technical assistance to help make markets in developing countries ready and able to implement low-carbon technologies. It's not just about funding individual projects, but about really <u>coming up with</u> more meaningful ways to partner around spreading technology our planet urgently needs. Industrialized countries cannot <u>leave developing ones to</u> "do this <u>on their own</u>".

**11** Some of the steel and cement (which is also used to make concrete) businesses working in developing countries are multinational companies which are bringing decarbonizing technologies into their operations from abroad. This is a good thing.

**12** But there are also local companies—including within the supply chains of these multinationals—which need to be involved in order to make decarbonization succeed.

**13** In India, for example, more than half of the steel manufacturing industry is small and medium-sized enterprises without the same access to these technologies. Does this local market currently have the technical **capacity** to adopt and service new hydrogen fuel **installations**, for example?

**14** Unfortunately, the answer is: Not really.

**15** In many cases, these local companies will likely <u>be unaware of</u> the need to actually change their practices to move towards something that's low-carbon—let alone how to do this and what technology options exist to help them. The speed of change needed means that the world cannot wait for them to do this alone.

**16** Governments everywhere have a role to play here, in ensuring that their policy frameworks drive decarbonization, promote the right technologies and prevent the **proliferation** of production processes that aren't low-carbon.

**17** Imagine: If construction products are in demand in a developing country and they're not already or sufficiently available on the market, a company or investor may see an opportunity to set up a new business—and if **stringent** regulations aren't in place, they might do this using outdated technology with higher emissions.

**18** Decarbonization is not the **mandate** of small steel and cement manufacturers, as participants noted in the pre-COP27 Asia consultation, or their area of expertise.

**19** It is an area that requires **collaboration** across different sectors—including to get better and more detailed data, and measurement, reporting and **verification** frameworks on emissions that can help guide government, and industry, decision-making.

**20** Steel and cement companies might often be seen by some of the public as "bad guys". Globally, these sectors do currently contribute about 50% of industrial greenhouse gas emissions.

**21** But they produce essential materials to build our houses, schools and cities and are needed for our growing communities. The demand should not be to stop production today, but to make it low-carbon today.

**22** Without more meaningful global partnerships on industrial decarbonization, there's a big risk that we won't be able to deliver on our climate commitments. We cannot afford this.

**23** Countries and industries globally need to move all together towards the same climate goals at the same time. Cooperation—including on policy, infrastructure development, and technology—will be key to doing this.

### New Words

**consultation** /ˌkɒnslˈteɪʃn/ *n.*
the act of discussing sth with sb or with a group of people before making a decision about it 咨询；商讨

**intensive** /ɪnˈtensɪv/ *adj.*
involving a lot of work or activity done in a short time 短时间内集中紧张进行的；密集的

**concrete** /ˈkɒŋkriːt/ *adj.*

based on facts, not on ideas or guesses 确实的，具体的（而非想象或猜测的）

**commitment** /kəˈmɪtmənt/ *n.*

a promise to do sth or to behave in a particular way; a promise to support sb/sth; the fact of committing yourself 承诺；许诺

**sufficient** /səˈfɪʃnt/ *adj.*

enough for a particular purpose; as much as you need 足够的；充足的

**boundary** /ˈbaʊndri/ *n.*

real or imagined line that marks the limits or edges of sth and separates it from other things or places; a dividing line 边界；分界线

**enforce** /ɪnˈfɔːs/ *v.*

to make sure that people obey a particular law or rule 强制执行，强行实施（法律或规定）

**critical** /ˈkrɪtɪkl/ *adj.*

expressing disapproval of sb/sth and saying what you think is bad about them 批评的；批判性的；挑剔的

**deliver** /dɪˈlɪvə(r)/ *v.*

to do what you promised to do or what you are expected to do; to produce or provide what people expect you to 履行诺言；兑现

**ensure** /ɪnˈʃʊə(r)/ *v.*

to make sure that sth happens or is definite 保证；确保

**reliable** /rɪˈlaɪəbl/ *adj.*

that is likely to be correct or true 真实可信的；可靠的

**infrastructure** /ˈɪnfrəstrʌktʃə(r)/ *n.*

the basic systems and services that are necessary for a country or an organization to run smoothly, for example buildings, transport and water and power supplies （国家或机构的）基础设施，基础建设

**facilitate** /fəˈsɪlɪteɪt/ *v.*

to make an action or a process possible or easier 促进；使便利

**implement** /ˈɪmplɪment/ *v.*

to make sth that has been officially decided start to happen or be used 执行；实施

**capacity** /kəˈpæsəti/ *n.*

the ability to understand or to do sth 领悟（或理解、办事）能力

**installation** /ˌɪnstəˈleɪʃn/ *n.*

a piece of equipment or machinery that has been fixed in position so that it can be used 安装的设备（或机器）

**proliferation** /prəˌlɪfəˈreɪʃn/ *n.*

the sudden increase in the number or amount of sth; a large number of a particular thing 激增；增殖

**stringent** /ˈstrɪndʒənt/ *adj.*

laws, rules, or conditions are very severe or are strictly controlled （法律、规定或条件）严格的

**mandate** /ˈmændeɪt/ *n.*

an official order given to sb to perform a particular task 委托书；授权令

**collaboration** /kəˌlæbəˈreɪʃ(ə)n/ *n.*

the act of working with another person or group of people to create or produce sth 合作；协作

**verification** /ˌverɪfɪˈkeɪʃn/ *n.*

the act of showing or checking that something is true or accurate 证实；核实

## Phrases and Expressions

**put sb/sth at risk:** to endanger someone or something 将……置于危险中

**speed up:** when a process or activity speeds up or when something speeds it up, it happens at a faster rate (过程、活动) 加速；使 (过程、活动) 加速

**come up with:** to devise or produce something 想起；提出

**leave sth/sb:** to go away from a place without taking sth/sb with you 忘了带；丢下

**on one's own:** if you do something on your own, you do it without any help from other people 独自地；自主地

**be unaware of:** not knowing or realizing that sth is happening or that sth exists 不知道；没意识到

**in demand:** wanted by many people; popular 受欢迎的；有需要

**set up:** if you set up somewhere or set yourself up somewhere, you establish yourself in a new business or new area 开办；开创事业

**in place:** in the correct position or arrangement 在合适的位置

**see as:** to consider or imagine that something is or might be a particular type of thing 把……看作

## Exercises

**I Understanding the Text**

1. Recognize the sequence in which the passage presents its most important information on the importance of joint efforts to decarbonize industry. Fill in the form with the correct choice from the following list.

① Critical questions urgently needed to address

② The background information and thesis statement

③ Conclusion

④ Problems faced by some local companies, such as small steel and cement manufacturers, and their solutions

⑤ Problems faced by many developing countries and their solutions

| Parts | Paragraphs | Main ideas |
|---|---|---|
| Part One | Paras.1-3 | |
| Part Two | Paras.4-5 | |
| Part Three | Paras.6-11 | |
| Part Four | Paras.12-21 | |
| Part Five | Paras.22-23 | |

2. Focus on the sentences in the text that express problems, solutions and evaluations. Pay attention to the sentence patterns and fill in key information.

**Problem 1** Currently, many developing country governments _____ on the emissions of their different industries and how they compare internationally.

**Problem 2** Relatively _____ so far in the way of infrastructure to _____ for industrial decarbonization.

**Problem 3** _____ is largely concentrated within industrialized countries and large multinational companies.

**Solution 1** We _____ in research and development into new technologies.

**Solution 2** We also _____ that exist but are not yet widely available.

**Solution 3** This likely _____ to help make markets in developing countries ready and able to implement low-carbon technologies.

*Further practice: Analyze Part 4 in the problem-solution pattern.*

**Problem  1**

**Solution  1**

**Problem  2**

**Solution  2**

**II Focusing on Language in Context**

1. Key Words & Expressions

A. Fill in the blanks with the words given below. Change the form where necessary. Each word can be used only once.

**consultation  deliver  concrete  stringent  implement  commitment  facilitate  reliable  enforce  collaboration**

① He has promised to finish the job by June and I am sure he will_____.

② Our information comes from a _____ source.

③ The decision was taken after close _____ with local residents.

④ The government worked in close _____ with teachers on the new curriculum.

⑤ The new trade agreement should_____ more rapid economic growth.

⑥ The hospital has a_____ to provide the best possible medical car.

⑦ It's the job of the police to_____the law.

⑧ It is easier to think in _____ terms rather than in the abstract.

⑨ He announced that there would be more_____ controls on the possession of weapons.

⑩ The government promised to _____a new system to control financial loan institutions.

B. Fill in the blanks with the phrases given below. Change the form where necessary.

**put..at risk; speed up; come up with; be unaware of; in demand; set up; in place; see… as; on one's own**

① Everyone needs to be _____ before the show begins.

② Rather than just being dejected by failure, try to _____ it _____an opportunity to improve yourself and learn from your mistakes.

③ Recently researchers have _____ a new theory.

④ The mayor's plan offers incentives to firms _____ in lower Manhattan.

⑤ It was so irresponsible of him to leave his children in the car all alone. Didn't he realize that he was _____ them _____?

⑥ I had already taken steps to _____ a solution to the problem.

⑦ He _____ completely _____ the whole affair.

⑧ She wanted me to strike out _____, buy a business.

⑨ Well-qualified young people with experience in marketing are very much _____ at the moment.

2. Usage

A. Complete the following sentences in the text by using "it" as a formative subject.

① During these meetings, _____the pace of progress so far is too slow and that puts us at real risk of not meeting global climate commitments. （显然的）

② _____for industrialized countries to lower emissions within their boundaries and enforce restrictions for products entering their markets. （这确实是不够的）

B. Complete the following sentences in the text by using "let alone".

① In many cases, these local companies will likely be unaware of the need to actually change their practices to move towards something that's low-carbon—_____
_____. （更不用说如何做到这一点，以及有哪些技术选择可以帮助他们）

② It is incredible that the 12-year-old managed to even reach the pedals, _____
_____. （更不用说开车）

③ There isn't enough room for us, _____. （更不用说任何客人了）
④ I barely had time to take a shower, _____. （更不用说在家做饭了）

**III Translation**

Translate the following passage into English.

中国是发展中国家中的大国，其工业化正在快速发展，环境问题也变得日益严峻，因此环境保护被国家视为一项基本国策。近年来，国家采取了很多措施来加强环境治理，如建立了世界著名的生态工程"三北防护林工程"（the Three-North Shelter Forest Program）。此外，中国也在大力发展自然保护区，颁布了《环境保护法》（The Law on Environmental Protection），加强环保意识和环保教育。目前，环境治理已取得明显成效，大部分城市环境和农业生态环境得到了很大改善，工业污染防治能力也大大提高。

**IV Pair Work**

Talk about classic examples of industry pollution in the world and the lessons that can be drawn from them.

**TEXT B**

微课

## Road to Greater Green Consumption

Li Hongyang

(A new plan unveils concrete steps for institutions and individuals alike.)

1　China will improve green power consumption in leading private and State-owned companies involved in driving economic growth, according to an implementation plan for promoting green consumption released last month.

2　Green power refers to power **generated** from renewable resources.

3　Local authorities should set a minimum level of green power consumption for energy-intensive enterprises to ensure the use of green energy, according to the plan released by the

National Development and Reform Commission, the country's top economic planner.

4    Local authorities are to **instruct** power grid operators to let users know when green power is available and encourage them to shift consumption to these periods.

5    On the **premise** that enough power is available, priority should be given to users with high levels of green power consumption.

6    The country will also continue to promote the development of smart photovoltaics and increase green power consumption by residents.

7    The plan proposed measures to develop green consumption in other key industries and sectors, including food, clothing, residential buildings and transportation.

8    "Green consumption requires that people practice green, low carbon philosophy throughout the process of consumption. <u>Promoting green consumption is a **profound** change, important to boosting high-quality development</u>," said Ha Zengyou, head of the NDRC's department of employment, income distribution and consumption, at a news conference last month.

9    Under the plan, green consumption patterns will become a conscious choice, and green and low-carbon products will become **mainstream** by 2030. A sound green consumption system, policies, and institutional mechanisms will also be established.

10    <u>To achieve this goal, the plan seeks to improve standards of agricultural production, storage, transportation and processing to reduce waste</u>. Authorities should strengthen supervision of food producers and <u>catering</u> operators <u>to</u> avoid unnecessary waste.

11    The plan encourages green consumption in the clothing industry by promoting the use of energy-saving printing, dyeing and waste fiber recycling equipment. Businesses, institutions and schools should encourage the wearing of green and low-carbon certified uniforms.

12    It also seeks to promote the large-scale development of green and low-carbon buildings. <u>For example, construction companies should add energy conservation and environmental protection into plans for the renovation of neighborhoods.</u>

13    Wang Bin, an official from the Ministry of Commerce, told the news conference that for the first three quarters of last year, about 25 percent more restaurants offered smaller dishes than normal on food delivery platforms to reduce waste. The number of people who bought **takeaways** without requesting disposable tableware exceeded 100 million.

14　During the Double Eleven (November 11) shopping festival period last year, sales volumes for energy-saving electric fans nearly tripled, while those for central air conditioners nearly doubled year-on-year.

15　In addition, the country will also **vigorously** promote new energy vehicles by removing restrictions on their purchase and building more charging, storage and hydrogen refueling infrastructure.

16　New energy vehicle sales have surged. Last year, China sold about 3.5 million new energy vehicles, 1.6 times more than the previous year. There were nearly 8 million new energy vehicles in the country, accounting for about half of the global number, said Wang Bin.

17　"The ministry has taken various measures to accelerate the purchase of green vehicles. For example, it promoted the elimination and renewal of diesel trucks, and supported withdrawal of high-emission vehicles from the market," Wang said.

18　The plan also included policies to support green consumption, among them proposals to improve the standard and certification system for green and low-carbon products and services.

19　Bo Yumin, an official from the State Administration for Market Regulation, said at the conference that the administration has issued certified product catalogs that include nearly 90 products.

20　"Green products refer to those that meet the requirements of resource conservation, environmental and consumer friendliness for their entire life cycle and process," Bo said.

21　"In terms of the selection of products for certification, priority is given to those that are closely related to food, clothing, housing, transportation, that have a great impact on health and the environment, and that have a certain market size and strong international demand," she said.

22　These include organic foods, textiles, automobiles and motorcycle tires, plastic products, washing products, construction materials, delivery packaging and electronic products.

23　Another policy offers targeted financial support to help improve the proportion of governmental purchases of green and low-carbon products, and encourages some areas to give appropriate subsidies or discounts on consumer goods including smart home appliances, green construction materials and other low-carbon products.

24　The plan further stated that the country will set up specialized green consumption guidance agencies and a national information platform, regularly publish lists of green and low-carbon products, and guides to facilitate purchasing by institutions and consumers.

## New Words

**unveil** /ˌʌnˈveɪl/ *v.*

to show or introduce a new plan, product, etc. to the public for the first time （首次）展示；推出

**alike** /əˈlaɪk/ *adv.*

used after you have referred to two people or groups, to mean "both" or "equally" 两者都；同样地

**generate** /ˈdʒenəreɪt/ *v.*

to produce or create sth 产生；引起

**instruct** /ɪnˈstrʌkt/ *v.*

to tell sb to do sth, especially in a formal or official way 指示；命令

**premise** /ˈpremɪs/ *n.*

a statement or an idea that forms the basis for a reasonable line of argument 前提；假定

**profound** /prəˈfaʊnd/ *adj.*

very great; felt or experienced very strongly 巨大的；深远的

**mainstream** /ˈmeɪnstriːm/ *adj.*

having, reflecting, or being compatible with the prevailing attitudes and values of a society or group 主流的

**takeaway** /ˈteɪkəweɪ/ *n.*

a meal that you buy at this type of restaurant 外卖的饭菜；外卖食物

**vigorous** /ˈvɪɡərəs/ *adj.*

very active, determined or full of energy 充满活力的；精力充沛的

## Phrases and Expressions

**seek to:** to try to do sth 试图；设法

**add...into...:** to include sth with sth else 把……加进去；包括

**take measures to:** to take an official action in order to achieve a particular aim 措施；方法

**refer to:** to describe or be connected to sth 涉及；与……相关

**in terms of:** used when you are referring to a particular aspect of sth 就……而言；在……方面

## Exercises

**I Comprehension Check**

① What is green power?

② What are the requirements of green consumption?

③ How does the concept of green consumption work in clothing industry?

④ How many people bought takeaways without requesting disposable tableware?

⑤ What do green products mean?

**II Translation**

Translate into Chinese the underlined sentences in the article.

① Local authorities should set a minimum level of green power consumption for energy-intensive enterprises to ensure the use of green energy.

② Promoting green consumption is a profound change, important to boosting high-quality development.

③ To achieve this goal, the plan seeks to improve standards of agricultural production, storage, transportation and processing to reduce waste.

④ For example, construction companies should add energy conservation and environmental protection into plans for the renovation of neighborhoods.

⑤ In addition, the country will also vigorously promote new energy vehicles by removing restrictions on their purchase and building more charging, storage and hydrogen refueling infrastructure.

**III Group Work**

What is green consumption? Illustrate it with a specific example in your daily life.

## TEXT C

**Chinese Recycling Company Sees Opportunities in Circular Economy**

CGTN

1    China's green development path has provided opportunities for its recycling industries, with some enterprises making achievements in the circular economy.

2    GEM Co. Ltd, a Shenzhen-based company that <u>specializes in</u> resource recycling, was awarded runner-up in the Award for Circular Economy Multinationals at the World Economic Forum in Davos, Switzerland, on Tuesday Beijing time.

3    The "Circulars", an initiative of the World Economic Forum (WEF) and the Forum of Young Global Leaders, is a circular economy awards program that **recognizes** individuals and organizations around the world that <u>make contributions to</u> the circular economy.

4    "We increasingly see China's globalizing enterprises playing a responsible role on the international stage in circular economy," said Klaus Schwab, WEF founder and chief executive.

5    GEM focuses on urban mines, including metal, used electronics and batteries. With China's booming new energy vehicle (NEV) market, it also recycles scrapped lithium batteries from NEVs by **extracting** the nickel, cobalt and other precious metals that are strategic resources.

6    Xu Kaihua, GEM chairman, believes that promoting the circular economy <u>is vital to</u> tackling the economic and environmental risks of resource exploitation.

7    "Turning waste into treasure is a sunrise industry that fits into China's green development. We aim to build a world-leading recycling enterprise," said Xu.

8    GEM is just one of the practitioners of China's green development concept. According to the report delivered at the 19th National Congress of the Communist Party of China in October, China will promote a sound economic structure that facilitates green, low-carbon, and circular development.

**9**   China aims to increase the output value of the resource recycling industry to 3 trillion yuan by 2020, according to an action plan to boost the recycling industry.

**10**   GEM has combined the recycling industry with green technology. It has applied for 1,200 core patents in the field of waste recycling and material recovery, including 52 Patent Cooperation Treaty (PCT) and foreign patents, among which over 20 core patents were **authorized** in Europe, the US and Japan.

**11**   The company has also promoted international cooperation in the circular economy. It cooperated with University of Oxford and the Oxford University Innovation Limited to research how to effectively produce pyrolysis oil from waste tires. GEM has also made investments in Europe and South Africa, and plans to build a China-African Circular Economy Industrial Park in South Africa.

**12**   Experts say China's recycling industry also sees opportunities in the global market.

"Many countries along the Belt and Road are still in the early stages of recycling their renewable resources. They have high demands for management experience and technical equipment, which will provide opportunities for China's maturing recycling enterprises," said Peng Xushu, deputy director of the Circular Economy Research Center at the Chinese Academy of Social Sciences.

## New Words

**recognize** /ˈrekəgnaɪz/ *v.*

to accept and approve of sb/sth officially （正式）认可；接受

**extract** /ˈekstrækt/ *v.*

to remove or obtain a substance from sth, for example by using an industrial or a chemical process 提取；提炼

**authorize** /ˈɔːθəraɪz/ *v.*

to give official permission for sth, or for sb to do sth 批准；授权

## Phrases and Expressions

**specialize in:** to become an expert in a particular area of work, study or business; to spend more time on one area of work, etc. than on others 专门研究（或从事）；专攻

**make contributions to:** to take an action that helps to cause or increase sth 贡献；促成作用

**be vital to:** is necessary or essential in order for sth to succeed or exist 必不可少的；对……极重要的

## Exercises

**Comprehension Check**

① What is the "Circulars"?

② What does Xu Kaihua think of circular economy?

③ What is the aim of China by 2020?

④ How many core patents were applied for by GEM?

⑤ Why do experts say China's recycling industry sees opportunities in the global market?

## Supplementary Reading

TEXT A

### China's Green Development in the New Era (1)

**Adjusting and improving the industrial structure**

1   China is committed to the philosophy of innovative, coordinated, green, open and shared development, and takes innovation-driven development as the driving force to create new momentum and build new strengths for economic development. China has placed rigid constraints on the exploitation of resources and the environment to promote comprehensive adjustment of the industrial structure, and strengthened regional cooperation to optimize the spatial configuration of industry. As a result, China's economy has registered a steady improvement in the quality of development while maintaining a reasonable pace of growth.

- Vigorously developing strategic emerging industries

2   China implements the innovation-driven development strategy. It takes scientific and technological innovation as the driving force and guarantee for adjustment of industrial structure and green and low-carbon transition of the economy and society and regards strategic emerging

industries as a key driver for economic development, reaping remarkable economic and social benefits as a result.

**3** China has intensified investment in scientific and technological innovation. The nation's gross domestic research and development (R&D) spending grew from RMB1.03 trillion in 2012 to more than RMB2.8 trillion in 2021. Its R&D spending intensity, or the expenditure on R&D as a percentage of its GDP, rose from 1.91 percent in 2012 to 2.44 percent in 2021, approaching the average level of the Organization for Economic Cooperation and Development (OECD) countries. Chinese enterprises' investment in R&D has continued to increase, accounting for more than 76 percent of the country's total R&D investment. By the end of 2021, China's energy conservation and environmental protection industry owned 49,000 valid invention patents, and the new energy industry held 60,000, 1.6 and 1.7 times more than in 2017. From 2011 to 2020, the number of patent applications filed by China for environment-related technology inventions was close to 60 percent of the world total, making it the most active country in environmental technology innovation.

**4** Emerging technologies have become the main props of China's economic development. Thanks to accelerated efforts to implement emerging technologies such as artificial intelligence (AI), big data, blockchain, and quantum communication, new products and business forms including intelligent terminals, telemedicine, and online education have been cultivated, and their role in boosting growth has continued to increase.

China's digital economy ranks second in the world. During the 13th Five-year Plan period (2016-2020), the average annual growth rate of the added value of information transmission, software and information technology services reached 21 percent. The internet, big data, AI, 5G and other emerging technologies are deeply integrated with traditional industries, facilitating the integration of advanced manufacturing with modern services. The value-added output of high-tech and equipment manufacturing in 2021 accounted for 15.1 and 32.4 percent of that of industries above designated size, up 5.7 and 4.2 percentage points from 2012 respectively. China is on the way to realize the transformation and upgrading from "made in China" to "intelligent manufacturing in China".

**5** China's green industries continue to grow. The renewable energy industry is growing rapidly, and China leads the world in the manufacture of clean energy generation facilities for wind and photovoltaic power. China produces more than 70 percent of the global total of polysilicon, wafers, cells and modules. The quality and efficiency of the energy-saving and environmental protection industries have continued to improve. China has developed a green technical equipment manufacturing system covering various sectors such as energy and water conservation, environmental protection, and renewable energy. The manufacturing and supply capacity of green technical equipment increases markedly while the cost keeps dropping. Technology in the fields of energy and water conservation equipment, pollution control, and environmental monitoring meets the highest international standards.New forms and models of business continue to grow, such as comprehensive energy services, contract-based energy and water management, third-party treatment of environmental pollution, and comprehensive carbon emissions management services. In 2021, the output value of China's energy conservation and environmental protection industries exceeded RMB8 trillion.Extensive pilot projects have been carried out at local level to explore methods and pathways to realize the value of eco-environmental products. New models of eco-friendly industry such as urban modern agriculture, leisure agriculture, eco-environmental tourism, forest healthcare, boutique homestays, and pastoral leisure complexes have witnessed rapid development.

• Taking well-ordered steps to develop resource-based industries

6  China continues to expand supply-side structural reform and reverse the extensive development model that relies heavily on resource consumption at the cost of high pollution and emissions. With environmental capacity as a rigid constraint, it has exerted tight control over the production capacity of energy-intensive industries and industries with high emissions or water consumption, in order to optimize its industrial structure.

7  Easing overcapacity and closing down outdated production facilities.While protecting industrial and supply chains, China has taken active and well-ordered steps to ease overcapacity and close down outdated production facilities. Measures have been taken to curb industries that over-exploit resources and cause environmental damage, such as steel, cement and electrolytic aluminum. A swap system has been introduced that allows producers to open equal or lower amounts of new capacity in return for closures elsewhere. During the 13th Five-year Plan period (2016-2020), China has removed more than 150 million tonnes of excess steel production capacity and 300 million tonnes of excess cement production capacity. Substandard steel products have been eliminated and almost all outdated production capacity in industries such as electrolytic aluminum and cement manufacturing has been removed.

8  China is resolved to stop the blind development of energy-intensive projects with high emissions and outdated production techniques. It has raised the entry threshold for some key industries in terms of land use, environmental protection, energy and water conservation, technology, and safety. A differentiated system has been introduced for energy-intensive industries, covering differentiated electricity pricing, tiered electricity pricing, and punitive electricity pricing.

For energy-intensive projects with high emissions and outdated production techniques, China applies a list-based management approach involving classification and dynamic monitoring. It resolutely investigates and punishes all projects that violate laws or regulations. In areas with problems of water shortage or overconsumption, restrictions are imposed on various types of new development zones and projects requiring high water consumption.

- Optimizing regional distribution of industries

9  Fully considering factors such as energy resources, environmental capacity, and market potential, China promotes the convergence of some industries in areas with more suitable conditions and greater potential for development. To expedite the formation of a modern and efficient industrial development configuration, it improves the distribution of productive forces and expands the division of industries and coordination across regions.

10  Working to bring about a rational distribution of raw material industries. China employs overall planning of resources such as coal and water and takes into consideration environmental capacity. Several modern coal chemical industry demonstration zones have been established in the central and western regions to pilot projects for technology upgrading in the coal chemical industry. A group of large-scale high-  quality petrochemical industry bases has been constructed in coastal areas to promote the safe, green, intensive, and efficient development of the industry.

11  Expanding the division of industries and cooperation across regions. China is seeking to establish and improve a benefit-sharing mechanism by employing the comparative strengths of every region, each relying on its own resources and environmental advantages, and on the foundations of industrial development. Multi-type and multi-mechanism industrial division and coordination have been strengthened, along with cooperation between the east and the central and western regions, creating a framework of coordination, complementarity of strengths, and common development.Transferring industries and cooperation across regions are measures that help to break through the environmental and resource constraints that stifle industrial development. They also make room for the development of high-tech industries in the eastern region and propel the industrialization and urbanization process of underdeveloped areas in the central and western regions, improving the balance and strengthening the coordination of regional development.

## TEXT B

## China's Green Development in the New Era (2)

**Extensive application of green production methods**

1  China has accelerated the building of a green, circular, and low-carbon economy. It practices green production methods, promotes the energy revolution, the economical and intensive use of resources, and cleaner production, and pursues synergy in the reduction of pollution and carbon emissions. All these efforts have contributed to the coordinated development and balanced progress of the economy, society, and environmental protection.

- Promoting the green transformation of traditional industries

2  In order to build a green, circular, and low-carbon production system, China has integrated the concept of green development into the entire life cycles of industry, agriculture and the service sector. To conserve energy, reduce emissions, raise efficiency, and facilitate the comprehensive green transformation of traditional industries, China has encouraged innovations in technology, models, and standards.

3  Promoting the green development of industry. China is committed to establishing a green manufacturing system, and creating green factories, green industrial parks, green supply chains, and green product evaluation standards. In order to accelerate the building of green industrial chains and supply chains, China provides guidance for enterprises to achieve innovations in the design of green products and adopt green, low-carbon and eco-friendly processes and equipment, and optimizes the spatial layout of enterprises, industries and infrastructure in industrial parks. Following the principles of "coupling of industries, extended responsibility of enterprises, and circular use of resources", it has promoted the transformation of industrial parks, circular combination of industries and circular production in enterprises. China has transformed its major industries to achieve clean production, and carried out comprehensive inspections of clean production. It has promoted digital transformation across the board.The digital control rate of key processes in key areas increased from 24.6 percent in 2012 to 55.3 percent in 2021, and the penetration rate of digital R&D and design tools increased from 48.8 percent to 74.7 percent in the same period. By the end of 2021, China hosted a total of 2,783 green factories, 223 green industrial parks, and 296 green supply chain management enterprises. The manufacturing sector has been significantly upgraded for green production.

4  Transforming the production methods of agriculture. China has created new systems and mechanisms for the green development of agriculture, expanded the functions of agriculture,

explored the diversified rural values, and strengthened the protection and efficient use of agricultural resources. It has gradually improved the farmland protection system and the system of fallowing and crop rotation, put permanent basic cropland under special protection, and thereby made initial progress in containing the decline in the size of farmland. It has steadily advanced the conservation of chernozem soil.The quality of farmland has been upgraded steadily throughout the country.Measures have been taken to save water for agricultural irrigation and reduce the volume of chemical fertilizers and pesticides used by targeting higher efficiency. In 2021, the irrigation efficiency was raised to 0.568. China has developed a circular agricultural economy by promoting circular agricultural production modes—integrating planting and breeding with processing, farming and animal husbandry with fishing, and production and processing with marketing. It has increased the utilization of agricultural waste as a resource. It has taken a coordinated approach to promoting green and organic agricultural products, products with quality certifications and those with geographical indications, cultivating new breeds, improving product quality, fostering agricultural brands and standardizing agricultural production.China has implemented programs to protect agricultural products with geographical indications. There are now 60,000 types of green food and organic agricultural products across the country. The quality and safety standards of agricultural products have been steadily upgraded. The supply of high-quality agricultural products has increased significantly, which has effectively contributed to the upgrading of the whole industry, and generated higher incomes for farmers.

5　Advancing the green transformation of the service sector. China has actively cultivated green firms of business circulation, and launched a campaign to create green shopping malls. Nationwide, a total of 592 green shopping malls had been built by the end of 2021. China has continued to improve the energy efficiency of the information service industry, with some world-leading green data centers. To accelerate  the reduction, standardization and recycling of express delivery packages, it has upgraded and improved the express delivery packaging standard system. To promote the green development of e-commerce enterprises, it has given guidance for producers and consumers to use renewable and degradable express delivery packages. By the end of 2021, 80.5 percent of e-commerce parcels were free of secondary packaging, all express delivery packages were sealed with thinner (45mm) tape, and all transit bags used in the sector were renewable.

6　China has promoted the green development of the convention and exhibition industry by formulating green standards and facilitating the repeated use of facilities. China has significantly reduced paper usage by introducing electronic railway tickets nationwide and encouraging electronic invoicing. In the catering industry, disposable tableware is being phased out. Guest houses and hotels have been encouraged not to offer disposable items as part of their services.

- Promoting green and low-carbon energy

7   China applies the principle of building the new before discarding the old in a well-planned way. With growing capacity to ensure energy supply, it has moved faster to build a new energy system. The proportion of clean energy sources has increased significantly. Success has been achieved in the green and low-carbon transformation of the country's energy mix.

8   Vigorously developing non-fossil energy. China has made rapid progress in building large-scale wind and photovoltaic power stations on infertile and rocky terrain and in deserts. It has steadily developed offshore wind farms, actively promoted rooftop photovoltaic power generation in urban and rural areas, and encouraged distributed wind power generation in rural areas. China has built a structured matrix of large hydropower stations in the basins of major rivers, especially those in the southwest. In accordance with local conditions, it has developed solar, biomass, geothermal and ocean energy, and power generation through urban solid waste incineration. It has developed nuclear power in a safe and orderly manner. Committed to innovation-driven development, China has worked on developing hydrogen energy. It has accelerated the construction of a new power system to adapt to the steady increase in the proportion of new energy. To promote the efficient use of renewable energy, it has carried out appraisals of relevant parties' performance in meeting the set goals for consumption of power generated from renewable energy. The proportion of clean energy sources in total energy consumption increased from 14.5 percent in 2012 to 25.5 percent by the end of 2021, and the proportion of coal decreased from 68.5 percent to 56 percent over the same period. The installed capacity of renewable energy was more than one billion kilowatts, accounting for 44.8 percent of China's overall installed capacity. The installed capacity of hydropower, wind power, and photovoltaic power each exceeded 300 million kilowatts, all ranking the highest in the world.

9   Advancing the clean and efficient use of fossil energy. To promote the clean and low-carbon development of coal-fired power, China has upgraded coal-fired power plants to conserve resources, reduce carbon emissions and make their operation more flexible, and transformed heating facilities. It has implemented stricter energy-saving standards for newly-installed coal-fired generating units. The efficiency and pollutant control levels of these units are on par with the most advanced international standards. China has promoted clean end-use energy by replacing coal with natural gas, electricity, and renewable energy. It has actively supported clean heating in winter in northern China. It has made the use of natural gas more efficient in urban areas, as well as in industrial fuel, power generation, and transport, and promoted natural gas combined cooling, heating, and power (CCHP). It has launched a campaign to upgrade the quality of refined oil products. In less than 10 years China achieved the upgrading that took developed countries 30-plus

years, and its refined oil products are now of the best quality by international standards. As a result, vehicle pollutant discharge has been effectively reduced.

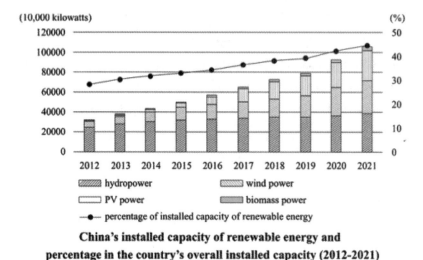

**China's installed capacity of renewable energy and percentage in the country's overall installed capacity (2012-2021)**

- Building a green transport network

**10**  The transport sector is one that consumes a large amount of energy and generates significant pollutant and greenhouse gas emissions. This is an area that deserves more attention in the pursuit of green development. By upgrading the energy efficiency of transport equipment, China has accelerated the building of a green transport network, with optimizing the structure of energy consumption and improving the efficiency of organization as its priorities, so that transport will be more eco-friendly, and travel will be more low-carbon.

**11**  The Beijing Winter Olympic Games opened on February 4, 2022. These games were different from their predecessors in that green electricity was used in all the 26 venues in the three competition zones — the first time in the history of the Olympic Games that all the venues had been powered by green electricity. Green electricity served all the purposes of the Beijing Winter Olympic Games — venue lighting, ice surface maintenance, production of artificial snow, TV broadcasting, timekeeping, security and logistical support. China put into action the concept of sustainable development it advocated when bidding for the Games. To supply the Games with green electricity, China built a large number of wind and PV power projects in Beijing, Zhangjiakou and other regions, and launched the Zhangjiakou-Beijing flexible HVDC pilot project, to transmit clean electricity to the Games venues. This not only met the demand of the Games for lighting, operations, transport and other purposes, but also raised by a substantial margin the share of clean energy consumption in Beijing and surrounding areas. Seizing the opportunities presented by hosting the green Winter Olympic Games, China has realized the large-scale transmission, grid-connection, and uptake of clean energy, accumulating valuable practical experience for the further development of clean energy, and demonstrating its confidence and determination in achieving the goals of carbon emissions peaking and carbon neutrality.

12   Optimizing the structure of transport. China has accelerated the construction of special railway lines, promoted the shift of freight transport from road to railway and waterway, and encouraged intermodal transport. In 2021, the railway and waterway freight volume accounted for 24.56 percent of the total in China, an increase of 3.85 percentage points over 2012. China has also emphasized the strategy of giving priority to urban public transport. By the end of 2021, there were 275 urban rail transit lines in operation in 51 cities, with a total track length of more than 8,700 kilometers. The length of exclusive bus lanes increased from 5,256 kilometers in 2012 to 18,264 kilometers in 2021.

13   Promoting the green transformation of transport vehicles. China has vigorously promoted the use of new-energy vehicles in public transport, taxi service, environmental sanitation, logistics, distribution, civil aviation, airports, and Party and government institutions. By the end of 2021, the number of China's registered new energy vehicles had reached 7.84 million, accounting for about half of the global figure. There were 508,900 new energy buses, accounting for 71.7 percent of the total number of buses in China.There were 207,800 new energy taxis. China has continued the green transformation of mobile railway equipment. The proportion of internal combustion locomotives decreased from 51 percent in 2012 to 36 percent in 2021.China has also updated the pollutant discharge standards for motor vehicles, promoted the use of liquefied natural gas (LNG) powered boats and transformation of shore power facilities, and accelerated the transformation or elimination of obsolete vehicles and boats. Since 2012, more than 30 million yellow-label vehicles with high emissions have been eliminated, and 47,100 obsolete inland river boats have been re-engineered or mothballed.

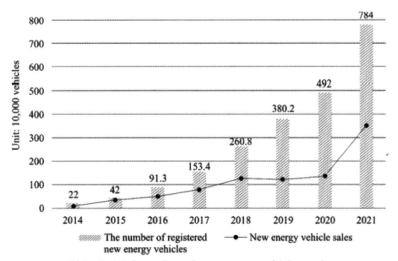

**China's total number of new energy vehicles and such vehicle sales (2014-2021)**

**14** Upgrading transport infrastructure for green development. China has initiated a special program for the construction of green highways, and the recycling of waste road surface materials. By the end of 2021, more than 95 percent of the waste materials from expressways and 80 percent of the waste materials from national and provincial highways had been recycled. China has steadily improved afforestation along its roads. Green belts have been built along 570,000 kilometers of its trunk roads, about 200,000 kilometers more than in 2012. China has continued the electrification of its railways, with the proportion of electric railways increasing from 52.3 percent in 2012 to 73.3 percent in 2021. It has also built more green port and road transport support facilities. By the end of 2021, five types of shore power facilities had been built in 75 percent of the specialized berths of major ports, and 13,374 charging piles had been built in expressway service areas — the highest number in the world.

- Promoting the economical and intensive use of resources

**15** As a country with a great demand for resources, China has accelerated the fundamental change in the way resources are utilized. To make a major contribution to the sustainable development of global resources and the environment, and to ensure a happy life for the people today as well as sufficient resources to meet the needs of future generations, China tries to obtain the maximum social and economic benefits at a minimum cost in resources and the environment.

**16** Improving the efficiency of energy use. China is exercising better control over the amount and intensity of energy consumption, particularly the consumption of fossil fuels. It has vigorously promoted technical, managerial, and structural energy conservation, to constantly improve the efficiency of energy use. It has initiated campaigns for all industrial enterprises, especially the big consumers of energy, to save energy, reduce carbon emissions, and improve the efficiency of energy use. The "forerunners" have been encouraged to play an exemplary role for other enterprises. China has organized the transformation of energy-intensive industries such as steel, power generation, and chemicals, to help them save energy and reduce carbon emissions. It has also strengthened the energy-saving management of key energy consumers, to enable large and medium-sized enterprises in key industries to reach advanced international levels in energy efficiency. Since 2012, China's average annual economic growth of 6.6 percent has been supported by an average annual growth of 3 percent in energy consumption, and the energy consumption per RMB10,000 of GDP in 2021 was 26.4 percent lower than in 2012.

**17** Improving the efficiency of water utilization. China has imposed increasingly rigid constraints on water use. Industrial and urban configurations are determined scientifically in accordance with water availability. China has launched nationwide water-saving campaigns to control the total amount and intensity of water consumption. It has upgraded water-saving technologies for industries with high water consumption, and promoted highly water-efficient irrigation for agriculture. It has advocated the building of water-saving cities, established a water efficiency labeling system, introduced certification standards for water conservation products, and promoted the use of water-saving products and appliances. The comprehensive per capita water consumption in cities is falling steadily. China has also incorporated unconventional water sources, such as reclaimed water, desalinated seawater, collected rainwater, brackish water, and mine water, into the unified allocation of

water resources, which has effectively eased the strain on demand in areas with a shortage of water. Water consumption per RMB10,000 of GDP in 2021 was 45 percent lower than in 2012.

**18** Strengthening the economical and intensive use of land. China has improved the standards for urban and rural land use. The designation, standards and approval of land use for all kinds of construction projects are strictly controlled, and the economical and intensive use of land in the construction of transport, energy, and water infrastructure is encouraged. China has strengthened the management of rural land, and promoted the economical and intensive use of rural land for collective construction projects. It has also established mechanisms for coordinating the use of existing land resources and made the arrangements for additional resources, and for recovering idle land, in order to put all existing land resources to good use. From 2012 to 2021, the area of land designated for construction projects per unit of GDP decreased by 40.85 percent.

**19** Making scientific use of marine resources. China has strictly controlled land reclamation from the sea. It has prohibited all coastal reclamation activities except those for major national projects, and dealt with problems left over from history in this regard with different approaches. It has established a control system to retain natural shorelines, and carried out classified protection and economical utilization of them. It has strictly protected uninhabited islands at sea and minimized their development and utilization.

**20** Ensuring the comprehensive use of resources. China has advocated the construction of green mines, promoted green exploration and exploitation, and worked to increase the recovery rate, processing recovery rate, and multipurpose utilization rate of major mineral resources. A total of 1,101 state-level green mines have been built. China has selected a total of 100 pilot projects and 100 backbone enterprises to promote the comprehensive use of resources and started the construction of national demonstration bases for recovering mineral resources from city waste. It has also updated the waste material collection network, coordinated the recycling of waste resources, and improved the processing and utilization of renewable resources. In 2021, 385 million tonnes of nine renewable resources—waste iron and steel, copper, aluminum, lead, zinc, paper, plastic, rubber, and glass—were recycled for new purposes.

## TEXT C

### Artificial Intelligence for Recycling: AMP Robotics

1  AMP's artificial intelligence (AI) platform AMP NeuronTM uses cameras to scan mixed waste streams and identify the different materials. Neuron's deep learning capability allows for continuous improvement of the identification and categorisation of paper, plastics and metals, by colour, size, shape, brand and other traits.

2  AMP CortexTM is the body to AMP Neuron's brain. Cortex is a high-speed intelligent robotics system that performs the physical task of sorting, picking and placing material based on information fed by the 'eyes and brain' of AMP Neuron. Cortex can sort recyclables at a rate of 80 items per minute with an accuracy of up to 99%.

3  AMP Neuron encompasses the largest known real-world dataset of recyclable materials for machine learning, with the ability to classify more than 100 different categories and characteristics of recyclables across single-stream recycling, e-scrap and construction and demolition debris, and reaching an object recognition run rate of more than 10 billion items annually.

**Why it's an example of a circular economy**

4  Globally, AMP has estimated that more than USD 200 billion worth of recyclable materials goes unrecovered annually. The economics and efficiency of identifying and sorting paper, plastics, metals, and other recyclables from the waste stream creates a major challenge for material recovery. In recent years, the waste industry has also faced stricter international quality standards for contamination-free imports of recyclable materials, leaving the industry in search of cost-effective ways to meet these requirements.

5  AMP's technologies allow more recyclables to be captured from waste streams, producing a greater volume of high-purity secondary resources. In one recycling centre in Virginia that installed the technology, the volume of recycled material increased by 10%.

6  The job of material sorting, which is typically carried out by humans, is physically demanding, poses safety risks and is prone to human error. Having humans and robots working

side by side, can stabilise the workforce and create improvements in the quality of full-time jobs. In the Virginia plant, the capital cost of the technology was offset through a reduction in recruitment and training costs of hard-to-find temporary workers.

**AI-enabled recycling is scaling fast**

7    In a sign of the proliferation of AI-driven recycling, AMP partnered with Waste Connections to deploy 25 robots across its materials recovery facilities nationwide, the company's largest order to date.

8    AMP's reach extends into other areas of the circular economy; one of its first corporate partners is Keurig Dr Pepper (KDP). The companies worked together to equip AMP's robotics systems to properly identify and sort K-Cup coffee pods in recycling facilities. AMP's installation at Evergreen, one of the largest recyclers of PET bottles in the US, demonstrates the company's continuous market expansion with plastics reclaimers in addition to its infrastructure modernisation efforts with materials recovery facilities.

**Benefits**

9    Applying AI and automation to waste sorting offers opportunities to improve performance and reduce operating costs, increasing the value of the secondary resources and improving the economics of recycling.

10    The data that is captured by AMP's technology can be used to save time and lower costs across many areas of a materials recovery facility by:

- Avoiding costly rejections by validating contamination-free material for resale
- Eliminating error-prone manual monitoring processes
- Preventing system downtime by anticipating issues
- Spotting hazards that could cause physical risks to staff
- Identifying volume and material composition trends to optimise operations

### Writing

People can reduce, reuse or recycle waste instead of throwing it away. What's the importance of this practice? Illustrate your ideas with specific examples.

### Unit Project

Work in groups to search for circular economy examples in the world in various fields, such as food, fashion, cities, etc. Make presentations in class and share your ideas with your classmates.

# Unit 5

# Biogeochemical Cycles of Nature

❝ Mass cannot be created or destroyed", this is the law of conservation of mass. Unlike energy, which can be dissipated as heat when flowing through different systems, elements of mass are recycled, and the biogeochemical cycles describe the movement of these elements between biotic and abiotic factors. The term biogeochemical is derived from "bio" meaning biosphere, "geo" meaning the geological components and "chemical" meaning the elements that move through a cycle. The six most common elements are carbon, nitrogen, hydrogen, oxygen, phosphorus, and sulfur. Water, which contains hydrogen and oxygen, is essential for living organisms, which makes water cycle as one of the primary cycles we care about. However, there are some other key elements that keep our bodies running. For example, carbon is found in all organic macromolecules and is also a key component of fossil fuels. Nitrogen is needed for our DNA, RNA, and proteins and is critical to agriculture. Thus, this chapter introduces to you some basic information about the water, carbon and nitrogen cycles. Moreover, biogeochemical cycles do not exist alone. They interact with one another and form a complete cycle of life. The life cycle concept will also be introduced with examples including the life cycle of a plastic bottle.

Practical English for Ecological Environment

## Warm-up

**Task 1** Listen to the audio and take notes according to the clues given.

(1) The route of the carbon cycle: sun→energy→plant→animal→$CO_2$.

(2) What happens when carbon-based organisms like plants or animals get stuck in the earth?

(3) What happened since the Industrial Revolution?

(4) What are the ways to re-balance the ecosystem?

**Task 2** Listen to the audio again and fill in the blanks with the missing information. You can refer to the notes you have taken.

(1) Plants create a stored form of energy, in the form of _____ such as glucose and sucrose. The process is called _____.

(2) _____ are a byproduct released through waste when animals like us eat those plants our stomachs convert that food back into energy for our own growth.

(3) If plants die, they _____, and tiny microorganisms break down those _____ and again, release greenhouse gases as a byproduct.

(4) Carbon based organisms trapped in the earth are compressed under tons of pressure, and turned into carbon-based _____ like oil, coal or natural gas.

(5) Since the Industrial Revolution, humans have been pulling those fossil fuels out of the ground and burning them, activating the stored energy to make _____ and power _____.

(6) _____ is defined as a technique to solve a problem. Sustainable technologies are those whose _____ is equal to their _____. They do not create negative externalities, such as $CO_2$, in the present or the future.

**Task 3** Work in small groups. Discuss with your group members the following question.

Water has been important for people for thousands of years. Without water there would be no life on earth. Make a list of the use of water in our daily lives and discuss how it affects the water cycle.

# Reading Comprehension

TEXT A

## The Life Cycle of Plastic Water Bottles

1   A lot of us are switching to reusable products—including water bottles—to be sustainable, but there are situations where single-use bottles are more appropriate and necessary. These instances are where the plastic water bottles mainly apply, such as in an emergency **evacuation** in a natural disaster, where people would have to leave suddenly.

2   Alternatively, if there was no running water in a community because of a damaged water pipe or **contamination** of the water supply then the disposable plastic bottles come in handy.

3   Given that plastics are not going to be completely phased out of our lives for a while, particularly plastic water bottle items, let us have a look at the life cycle of a plastic water bottle. We shall undertake this task through thorough intel on the manufacturing, processing, **distribution**, consumption, **disposal**, and ultimately recycling of these plastic water bottles.

### How are plastic water bottles manufactured?

4   To create the plastic needed to make these plastic bottles, the **initial** step involves **extraction** of crude oil from the earth. The crude oil is essential in this manufacturing process as polyethylene terephthalate (PET) is the main component in producing plastics.

5   This crude oil is then shipped to a refinery, where it is broken down or refined into its components; fuel oil, gas, etc. In a factory the gas and oil are chemically bonded with the resulting molecules creating monomers. Monomers are then further chemically broken down and bonded into polyethylene terephthalate, or PET, the polymer that comprises most of these plastic water bottles.

6   At this point, the final products are tiny pellets. Usually, the pellets are sorted and grouped according to color to produce similarly colored bottles. The pellets **undergo** a process of melting and are later **injected** into molds to fashion the shape and design of the plastic bottle.

7   The factory ships out these finished bottles to respective bottling plants, which fill the bottles with water, label them, package them, and eventually have them **delivered** to stores for consumers.

**How are plastic water bottles consumed and disposed of?**

8   The plastic water bottles, fresh from the bottling company, then ships them to the grocery store, **vendors**, and other places. These are stations or **venues** where the water is sold or given away, being available to the consumer for buying and drinking.

9   After consumers buy them and the water is drunk, the plastic water bottle is then thrown in the garbage or in rare cases, recycled. Unfortunately, a vast majority of water bottles go to the landfill (73%—90% of them) where they can take up to 500-1000 years before **decomposing**.

10   As a consumer, you should strive to recycle these plastic water bottles, via recycling facilities, as opposed to ignorantly throwing them in trash bins.

**What happens to bottles that are recycled?**

11   The other 10%—23% of plastic water bottles that fortunately make it to a recycling facility undergo the melting process of pellets all over again to produce other plastic products, such as water bottles, shopping bags, and even carpeting items.

12   The following steps offer a guide to how this recycling process works.

13   Initially, everything that is going to be recycled is put on a disk screen for sorting and separation. The separation of plastic water bottles mainly relies on category and type.

14   The sorted plastic bottles are then collected into a large bale and transported to a different facility. This facility primarily deals with shredding, sorting, washing, and heating these plastic water bottles to create a resin. The final resin is eventually ground into pellets to create more material for plastic products.

15   Recycling plastic bottles is great from the standpoint of using less crude and oil, which most of us know is a finite resource. Therefore, while recycling plastic may be a bit complex, it is still a worthwhile **endeavor** until we come up with a better solution.

16   The good news is that there are **innovations** set in motion to create new single-use products that can break down faster, and use non-toxic components or finite resources.  Take edible water bottles for example! Some manufacturing companies are investing in these **alternative**

types and other forms of bio plastics, obtained from plant products, instead of extracting crude and oil.

**Benefits of recycling plastic water bottles**

17 • **There are low amounts of wastes in general.**
- The process and activity of recycling are simple, fun, and easy.
- Pollution rates lower due to low energy consumption.
- There are low greenhouse gas emissions in the environment.
- Saves on energy.
- Uses fewer resources meaning less need for mining of finite resources such as crude oil.
- Saves money as the cost of manufacturing and processing lowers significantly.
- Creates jobs for people in the collection, recycling, and processing industries.
- Maintains sustainability of resources such as the crude oil used for plastics manufacturing.

18  We all should <u>care about</u> looking after our environment. Some resources are finite and some of the materials we dispose of negatively affect the ecosystem. Our disregard for nature affects our oceans, the animals and fish in them, and ultimately the animals on land and ourselves. So, not only are we looking after the planet by recycling, but also the other living creatures and humans we share it with.

19  We need to do the best we can with what we have, and know, now. We can also vote with our wallets and support companies that are looking for ways to be environmentally responsible.

20  If you already recycle, well done, and <u>keep up</u> the good fight!

### New Words

**evacuation** [ɪˌvækjuˈeɪʃn] *n.*
the act of evacuating 撤离；疏散
**contamination** [kənˌtæmɪˈneɪʃən] *adj.*
the state of being contaminated; a substance that contaminates 污染；污染物

**distribution** [ˌdɪstrɪˈbjuːʃn] *n.*

the commercial activity of transporting and selling goods from a producer to a consumer 销售

**disposal** [dɪˈspəʊzl] *n.*

action of getting rid of sth 清除；处理；处置

**initial** [ɪˈnɪʃl] *adj.*

occurring at the beginning 最初的；开头的

**extraction** [ɪkˈstrækʃn] *n.*

the process of obtaining something from a mixture or compound by chemical or physical or mechanical means 抽出

**undergo** [ˌʌndəˈɡəʊ] *v.*

pass through 经历

**inject** [ɪnˈdʒekt] *vt.*

put (liquid) into (sb) with a special needle; fill (sth with a liquid, etc.) by injecting 注射；注入

**deliver** /dɪˈlɪvə(r)/ *v.*

bring to a destination, make a deliver 递送

**vendor** [ˈvendə(r)] *n.*

someone who promotes or exchanges goods or services for money 小贩；卖方；供货商

**venue** [ˈvenjuː] *n.*

the scene of any event or action (especially the place of a meeting) 集合地；会场，场所

**decompose** [ˌdiːkəmˈpəʊz] *v.*

break down 分解；腐烂

**endeavor** [ɪnˈdevə] *n.*

earnest and conscientious activity intended to do or accomplish something 努力

**innovation** [ˌɪnəˈveɪʃn] *n.*

the creation of something in the mind 创新；革新

**alternative** [ɔːlˈtɜːnətɪv] *adj.*

(of two or more things) that may be used, had, done, etc., instead of another 两者择一的；供替代的

## Phrases and Expressions

**come in handy**：turn out to be useful 迟早会有用；派上用场

**break down:** When a substance breaks down or when something breaks it down, a biological or chemical process causes it to separate into the substances which make it up. 分解

**be bonded with:** be closely related to 与……联系紧密

**comprised of**: to consist of particular parts, groups 包含；包括

**at this point**: at this time 此时；在这个时候

**strive to**: make great efforts to achieve or obtain something 表示尽力去做某事；竭尽全力去实现某个目标

**as opposed to**: used to compare two things and show that they are different from each other 与……相对；用于表示与另一事物相对或对比

**make it**: to succeed in getting somewhere in time for something or when this is difficult 成功；达到预期目标

**all over again**: If you do something all over again, you repeat it from the beginning. 表示再次从头开始做某事

**relies on**: to depend on something in order to continue to live or exist 依赖

**from the standpoint of**: form a point of view 从……的角度来看

**set in motion**: to start a process or series of events that will continue for some time 启动；使开始运行

**care about**: to think that something is important, so that you are interested in it, worried about it 关心；关注

**keep up**: to continue doing something 保持

### Exercises

**I Understanding the Text**

1. Recognize the sequence in which the passage presents its most important information on the life cycle of plastic bottles. Fill in the form with the correct choice form the following list.

① The recycling processes of water bottles and its benefits

② Keeping up the recycling to look after the environment

③ The ways water bottles are consumed and disposed of

④ The instances where the plastic water bottles mainly apply

⑤ The processes of manufacturing water bottles

| Parts | Paragraphs | Main ideas |
| --- | --- | --- |
| Part One | Paras.1-3 | |
| Part Two | Paras.4-7 | |
| Part Three | Paras.8-10 | |
| Part Four | Paras.11-17 | |
| Part Five | Paras.18-20 | |

2. Focus on the sentences that express processes and procedures. Pay attention to the sentence patterns and fill in key information.

**Step 1:** To create the plastic needed to make these plastic bottles, _____ from the earth.

**Step 2:** This crude oil is _____ , where it is _____; fuel oil, gas, etc.

**Step 3:** Monomers are _____ and bonded into polyethylene terephthalate, or PET, the polymer that _____.

**Step 4:** _____ are tiny pellets. Usually, the pellets are sorted and grouped according to color to produce similarly colored bottles.

**Step 5:** The pellets undergo a process of melting and are _____ of the plastic bottle.

**Step 6:** The factory ships out these finished bottles to respective bottling plants, which fill the bottles with water, _____, and eventually _____ to stores for consumers.

Further practice: Find out the markers of processes and procedures in Part 4.

the following steps

…

**II Focusing on Language in Context**

1. Key Words & Expressions

A. Fill in the blanks with the words given below. Change the form where necessary. Each word can be used only once.

**distribution  evacuation  contamination  disposal  initial  undergo  inject  decompose  endeavor  alterative**

① We have the _____ plans of having a picnic or taking a boat trip.

② We believe any level of _____ is unacceptable.

③ The _____ of the magazine is 2000.

④ Orders went out to prepare for the _____ of the city.

⑤ After the _____ cheers, the noise of the crowd began to die away as the group started playing.

⑥ You can apply heat to _____ organic compounds.

⑦ The _____ of rubbish is always a problem.

⑧ His medicine is a drug that can be _____ or taken by mouth.

⑨ Where there is no hope, there can be no _____.

⑩ This city _____ great changes.

B. Fill in the blanks with the phrases given below. Change the form where necessary.

**come in handy; break down; strive to; make it; rely on; set in motion; care about; from the standpoint of; all over again**

① The bacteria in the soil help to _____ dead plants and animals.

② You should always _____ achieve more, however well you have done before.

③ He's hoping to _____ big on TV.

④ _____ success, a good work ethic is no less important than capabilities.

⑤ The extra money _____.

⑥ The whole process started _____.

⑦ They also _____ minimizing energy use and reducing waste.

⑧ Lions _____ stealth when hunting.

⑨ His death _____ a train of events that led to the outbreak of war.

2. Usage

A. Complete the following sentences in the text by using attributive clause led by "where".

① A lot of us are switching to reusable products—including water bottles—to be sustainable, but there are situations _____. （一次性瓶子更合适，也更必要）

② These instances are where the plastic water bottles mainly apply, such as in an emergency evacuation in a natural disaster, _____. （这种情形下，人们会突然离开）

③ This crude oil is then shipped to a refinery, _____; fuel oil, gas, etc. （在那里它被分解或提炼成各种成分）

④ These are stations or venues _____, being available to the consumer for buying and drinking. （在那里有卖水或送水的）

⑤ Unfortunately, a vast majority of water bottles go to the landfill (73%—90% of them) _____. （在那里它们需要长达500—1000年才能分解）

B. Complete the following sentences in the text by using "given".

① _____ （考虑到）plastics are not going to be completely phased out of our lives for a while, particularly plastic water bottle items, let us have a look at the life cycle of a plastic water bottle.

② _____, （考虑到他的年龄）he's a remarkably fast runner.

③ _____, （考虑到他花了六个月的时间来做这件事）he hasn't made much progress.

④ At the top of the pyramid were a few colleges of advanced technology, but these were soon _____. （获得了独立大学的地位）

⑤ _____, （在某一特定时期）the value of shares will rise and fall.

**III Translation**

Translate the following passage into English.

4月22日是第53个世界地球日。如今，全世界每年以亿吨为单位产出的塑料垃圾，正使地球陷入巨大的塑料污染危机。从世界最高峰珠穆朗玛峰（Mount Qomolangma）的山巅，到地球海拔最低的马里亚纳海沟（Mariana Trench），甚至在人迹罕至的南极海域，都发现了塑料垃圾的踪迹。面对塑料污染这一全球共同关注的环境问题，中国正全力以赴，以实际行动彰显大国担当。从生产、使用、回收、处置、清理等各个环节持续推进塑料污染治理，我国塑料废弃物的环境污染得到了有效遏制。

**IV Pair Work**

Talk about the recyclable materials in your daily life.

TEXT B

## The Carbon Cycle

1　While most of the Earth's carbon can be found in the **geosphere**, carbon is found in all living things, soils, the ocean, and atmosphere. <u>Carbon is the primary building block of life, including DNA, proteins, sugars and fats</u>. One of the most important carbon compounds in the atmosphere is carbon dioxide ($CO_2$), while in rocks carbon is major component of **limestone**, coal, oil and gas. Carbon cycles through the atmosphere, biosphere, geosphere, and hydrosphere via processes that include photosynthesis, fire, the burning of fossil fuels, weathering, and volcanism. By understanding how human activities have altered the carbon cycle, we can explain many of the climate and ecosystem changes we are experiencing today, and why this rapid rate of change is largely unprecedented in the Earth's history.

**What is the carbon cycle?**

2　Carbon is transferred between the ocean, atmosphere, soil, and living things over time scales of hours to centuries. For example, photosynthesizing plants on land remove carbon dioxide directly from the atmosphere, and those carbon atoms become part of the structure of the plants. As plants are eaten by herbivores and herbivores are eaten by carnivores, carbon moves up the food web. Meanwhile, the **respiration** of plants, animals, and microbes returns carbon to the atmosphere as carbon dioxide ($CO_2$). When organisms die and decay carbon also returns to the atmosphere, or is integrated into soil along with some of their waste. The combustion of biomass during wildfires also release large amounts of carbon stored in plants back into the atmosphere.

(Global Change Infographic)

3　On longer timescales, significant amounts of carbon are transferred between rocks and the ocean and atmosphere, typically over thousands to millions of years. For example, the weathering of rocks removes carbon dioxide from the atmosphere. The resulting **sediments**, along with organic material, can be transported (**eroded**) from the land to enter the ocean where they sink to the bottom. This carbon from land, as well as carbon atoms in $CO_2$ absorbed by the ocean from the atmosphere, can become <u>incorporated into</u> calcium carbonate ($CaCO_3$) shells made by algae, plants, and animals. These shells become buried. As the successive layers of sediment are compressed and cemented they are turned into limestone rock. Over millions of years these carbon-bearing rocks can be exposed to sufficient heat and pressure to melt, causing them to release their carbon back into the atmosphere as carbon dioxide via volcanism. Some of these rocks will also be exposed at the surface of the Earth through mountain building and weathering, and the cycling begins again. Carbon from the mantle (see plate tectonics) is also released into the atmosphere as carbon dioxide through volcanic activity.

4   <u>Carbon is also transferred to rocks from the biosphere, via the formation of fossil fuels, which form over millions of years</u>. Fossil fuels <u>are derived from</u> the burial of photosynthetic organisms, including plants on land (which primarily forms coal) and plankton in the oceans (which primarily forms oil and natural gas). While buried, this carbon is removed from the carbon cycle for millions of years to hundreds of millions of years.

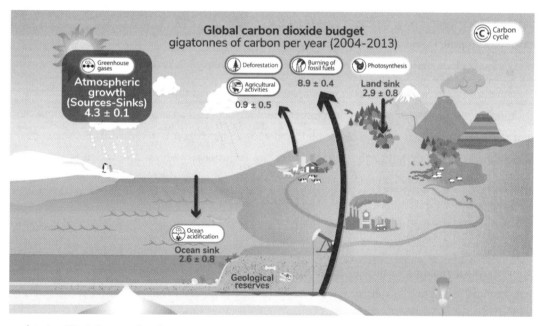

( A simplified diagram showing some of the ways carbon dioxide moves through the Earth system, and the overall increase in atmospheric carbon dioxide from 2004—2013 )

5   <u>Human activity, especially the burning of fossil fuels, has dramatically increased the exchange of carbon from the ground back into the atmosphere and oceans</u>. This return of carbon back into atmosphere as carbon dioxide is occurring at a rate that is hundreds to thousands of

times faster than it took to bury it, and much faster than it can be removed by the carbon cycle (for example, by weathering). Thus, the carbon dioxide released from the burning of fossil fuels is accumulating in the atmosphere, increasing average temperatures through the greenhouse effect, as well as **dissolving** in the ocean, causing ocean acidification.

6   The rate of exchange and the distribution of carbon in the Earth system is affected by various human activities and environmental phenomena, including:

·   The burning of fossil fuels, which rapidly releases carbon dioxide ($CO_2$), a greenhouse gas into the atmosphere, increasing average global temperatures and causing ocean acidification.

·   Agricultural activities that release carbon dioxide and methane ($CH_4$, a greenhouse gases) into the atmosphere. For example, methane is produced from the digestion of plant material by cows, and from the bacteria that thrive in rice fields. Carbon dioxide is released from the burning of fossil fuels to power farming equipment, from the mining of minerals and the making of fertilizer. The growing of crops and the raising of livestock also affects local productivity and biomass, and rates of photosynthesis, respiration, and decay of organic material.

·   Deforestation, which decreases rates of photosynthesis and thus how much carbon dioxide is captured by the growth of plants. When trees grow they take carbon dioxide out of the atmosphere and transfer it into their wood, leaves, bark and roots. The carbon is returned to the atmosphere when downed trees are left to rot, or if the trees are intentionally set on fire, which is a common means of deforestation. Thus, deforestation typically releases carbon dioxide, unless all the material is used for construction, or for paper products.

·   The extent of permafrost (soil that is frozen all year round), which contains methane ($CH_4$, a greenhouse gas). When temperatures remain cold all year-round organic material decays very slowly, and it remains in the soil. The melting of permafrost, which is happening as global temperatures increase, releases methane. The increasing temperatures also increase rates of decay, which further increases the amount of greenhouse gases in the atmosphere.

·   Over millions of years changes in the rate of sedimentation and rate of burial of organic matter alters the amount of carbon available for decay and how much carbon is stored in the rock record. For example, increased burial of dead plants and plankton decreases decay thereby increasing the rate of formation of fossil fuels.

·   Over millions of years, processes in the rock cycle can change carbon dioxide levels in the atmosphere. For example, the metamorphic reactions that occur under heat and pressure can release carbon dioxide. <u>In contrast, the weathering of rocks that occurs when carbon dioxide dissolves into rainwater to form carbonic acid ($H_2CO_3$) reduces the amount of carbon dioxide in the atmosphere.</u> Warming can increase these weathering reactions, but not at a rate that can offset the increase in carbon dioxide due to human activities.

- Geologic changes in the rate of volcanism, **driven** by plate tectonics, can dramatically **alter** the amount of carbon dioxide in the atmosphere, but on timescales much longer than human timescales, over millions of years.

**Earth system model about the carbon cycle**

7   The Earth system model below includes some of the processes and phenomena related to the carbon cycle. These processes operate at various rates and on different spatial and temporal scales. For example, carbon is transferred among plants and animals over relatively short time periods (hours-weeks), but the human extraction and burning of fossil fuels has altered the carbon cycle over decades. <u>Additionally, processes that include weathering and volcanism affect the carbon cycle over millions of years</u>. Can you think of additional cause and effect relationships between the parts of the carbon cycle and other processes in the Earth system?

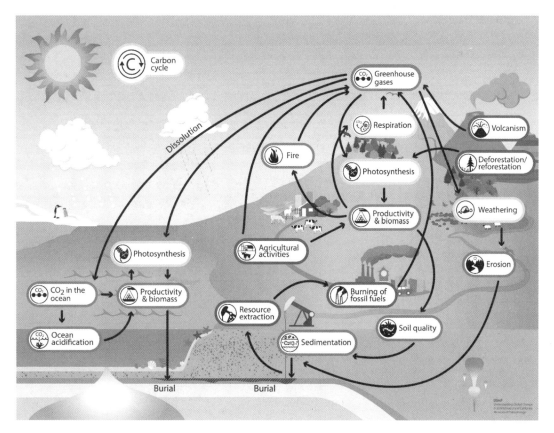

121

## New Words

**geosphere** [ˈdʒiːəʊsfɪə] *n.*
the solid part of the earth consisting of the crust and outer mantle 陆界
**limestone** [ˈlaɪmstəʊn] n.
a sedimentary rock consisting mainly of calcium that was deposited by the remains of marine animals 石灰石
**respiration** [ˌrespəˈreɪʃn] *n.*
processes that take place in the cells and tissues during which energy is released and carbon dioxide is produced and absorbed by the blood to be transported to the lungs 植物呼吸作用
**sediments** [ˈsedɪmənts] *n.*
matter that has been deposited by some natural process 沉淀物
**eroded** [ɪˈrəʊdɪd] *adj.*
worn away as by water or ice or wind 被侵蚀的
**dissolving** [dɪˈzɒlvɪŋ] *adj.*
the process of going into solution 毁灭性的；消溶的
**driven** [ˈdrɪvn] *adj.*
compelled forcibly by an outside agency 受到驱策的
**alter** [ˈɔːltə(r)] *v.*
cause to change; make different; cause a transformation 改变

## Phrases and Expressions

**incorporate into:** incorporate someone or something in(to) something 合并；吸收
**be derived from:** to come or develop from something 来源于

## Exercises

**I Comprehension Check**

① Where can carbon be found?

_____

② How can those carbon atoms become part of the structure of the plants?

_____

③ What has dramatically increased the exchange of carbon from the ground back into the atmosphere and oceans?

_____

④ How does deforestation affect the environment?

_____

⑤ What has altered the carbon cycle over decades?

**II Translation**

Translate into Chinese the underlined sentences in the article.

① Carbon is the primary building block of life, including DNA, proteins, sugars and fats.

② Carbon is also transferred to rocks from the biosphere, via the formation of fossil fuels, which form over millions of years.

③ Human activity, especially the burning of fossil fuels, has dramatically increased the exchange of carbon from the ground back into the atmosphere and oceans.

④ In contrast, the weathering of rocks that occurs when carbon dioxide dissolves into rainwater to form carbonic acid (H2CO3) reduces the amount of carbon dioxide in the atmosphere.

⑤ Additionally, processes that include weathering and volcanism affect the carbon cycle over millions of years.

**III Group Work**

Can you think of additional cause-and-effect relationships between the parts of the carbon cycle and other processes in the earth system?

**TEXT C**

**The Nitrogen Cycle in Aquariums—An Easy Guide for Beginners**

Alison Page

1  Unlike the natural habitats of most tropical fish species, where dangerous levels of

nitrogen-containing compounds are rare, the freshwater home aquarium is a closed environment in which overcrowding and overfeeding are common. That makes your fish tank <u>conducive to</u> excess ammonia and nitrification in the water, which is the most common cause of sickness, disease, and even death in aquarium fishes.

2   In this guide, we explain how the nitrogen cycle in the aquarium works, as well as giving you lots of helpful tips on how to keep your tank water **pristine** and safe for your fish.

**What is the nitrogen cycle?**

3   In nature, the nitrogen cycle describes the process where nitrogen moves from the air to plants, to animals to bacteria, and then back to air. That system works just fine and needs no human **intervention**. However, the cycle works differently in the enclosed environment of the aquarium.

4   In a fish tank, the process is a biochemical mechanism that sees the continual degradation of various nitrogenous compounds from ammonia to nitrite to nitrate. In the last phase of the cycle, nitrates are <u>taken up</u> by living plants in the tank and used as nutrients or removed from the water via partial water changes and your biological filtration system.

**Biological filtration system**

5   In the aquarium, it's up to the **hobbyist** to manage the nitrogen cycle effectively by using a biological filtration system. The biological filtration system contains sponges that act as a platform for the growth of beneficial bacteria. Those bacteria are vital to the nitrogen cycle process, as they work to <u>break down</u> the ammonia and nitrites produced by fish waste and decaying organic matter, keeping the water safe for your livestock.

6   A colony of bacteria in the filter system takes up to three months to become sufficiently established to be able to **convert** ammonia and nitrites to nitrate. For that reason, it's advisable to take time in stocking a new aquarium with a few small fish at a time to allow the biological filter to keep pace with the gradually increased bioload in the tank.

**Stages of the nitrogen cycle**

7   The nitrogen cycle is the process by which certain bacteria process harmful waste. There are three stages to the cycle:

- Waste Products Decay

8   The first stage in the nitrogen cycle is the decay of organic matter, such as uneaten food, dead plant leaves, dead organisms, and the waste produced by fish and invertebrates. As bacteria

cause these materials to break down, the metabolism of protein produces ammonia.

9   Ammonia is a colorless gas that is extremely **toxic** to fish, and even low levels in the water will cause oxygen **deprivation** and even burn the fishes' delicate **gills**. High ammonia levels in aquarium water usually occur because there are too many fish in the tank or the fish are overfed.

- Ammonia To Nitrite

10   In a balanced aquarium, nitrogen-fixing bacteria called Nitrosomonas consumes the ammonia, oxidizing it to create nitrite. Nitrite is also toxic, disrupting the oxygen-carrying capabilities of blood, but fish can generally <u>cope with</u> twice the quantity of nitrite in the water when compared to ammonia.

- Nitrite To Nitrate

11   In the third stage of the nitrogen cycle, bacteria called Nitrobacter process the nitrites, releasing a less toxic chemical called nitrate. Although nitrates are not toxic at low levels, if the concentration rises above 20 ppm, they can become dangerous to fish. In nature, nitrate converts aerobically into harmless nitrogen gas. That doesn't happen in the closed environment scenario of most aquariums, meaning that partial water changes are necessary to **dilute** the nitrate.

12   Live plants help to remove nitrates from the water in freshwater tanks. In a saltwater setup, live rock and deep sand beds can provide anaerobic areas where **denitrifying** bacteria process nitrates into nitrogen gas, which then evaporates harmlessly.

13   This aquarium nitrogen cycle diagram illustrates the process perfectly.

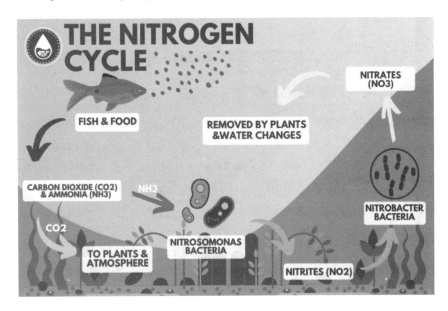

**Starting the nitrogen cycle**

14   So, you can see that the key to clean, healthy water that's free-from toxins and is safe for

your fish is to get the nitrogen cycle established and working efficiently. The process of establishing and maturing your biological filtration system is referred to in the hobby as "cycling."

15  In a new fish tank, you can cycle the aquarium with or without fish.

**Starting the nitrogen cycle with fish**

16  When you add fish to your new aquarium, they begin to produce waste and ammonia, and decaying fish food will also be added to the mix. Unfortunately, these fish often don't survive, earning them the nickname "suicide fish."

17  As a new filtration system does not contain any established Nitrosomonas bacteria colonies to consume the ammonia that the fish produce, levels of the toxin rise and ultimately **spike** until the bacteria population catches up.

18  You can see that phenomenon taking place quite easily as the water in the aquarium becomes temporarily cloudy. Once the bacteria take control of the situation by breaking down the ammonia, the water will clear, and the ammonia level drops.

**Nitrite spike**

19  As the Nitrosomonas bacteria consume the ammonia in the tank, nitrite is produced. As the number of bacteria increases, so does the level of nitrite. In response, the population of nitrite-eating Nitrobacter bacteria skyrockets due to the number of nutrients that are available to them.

20  Once that nitrite spike is reached, the levels of nitrite in the water will decrease, as the bacteria break down the chemical faster than it is being produced.

**Nitrate management**

21  The final end product of the nitrogen cycle is **nitrate**. While in low **concentrations** in the water, nitrate is not as dangerous to fish as ammonia or nitrites, although it can cause other problems in the tank, such as algae blooms.

22  To control the levels of nitrates in the water, you need to carry out partial water changes every week. The addition of live plants is also extremely effective in controlling nitrates as the plants extract the chemical from the **substrate** and from the water column to use as a nutrient-rich fertilizer.

### New Words

**pristine** [ˈprɪstiːn] *adj.*
completely free from dirt or contamination 未受损的；清新的
**intervention** [ˌɪntəˈvenʃn] *n.*

the act of intervening (as to mediate a dispute, etc.) 介入；干预

**hobbyist** [ˈhɒbiɪst] *n.*

a person who pursues an activity in their spare time for pleasure 业余爱好者

**convert** [kənˈvɜːt] *v.*

change the nature, purpose, or function of something （使）转变

**toxic** [ˈtɒksɪk] *adj.*

of or relating to or caused by a toxin or poison 有毒的

**deprivation** [ˌdeprɪˈveɪʃn] *n.*

the disadvantage that results from losing something 丧失

**gill** [ɡɪl] *n.*

respiratory organ of aquatic animals that breathe oxygen dissolved in water 鳃

**dilute** [daɪˈluːt] *vt.*

lessen the strength or flavor of a solution or mixture 稀释

**denitrifying** [ˈdenɪtrɪfaɪɪŋ] *adj.*

remove nitrogen from 脱硝的；脱氮的

**spike** [spaɪk] *v.*

to rise quickly and reach a high value 迅速升值；急剧增值

**nitrate** [ˈnaɪtreɪt] *n.*

any compound containing the nitrate group (such as a salt or ester of nitric acid) 硝酸盐

**concentration** [ˌkɒnsenˈtreɪʃən] *n.*

number of molecules of a substance in a given volume 浓度

## Phrases and Expressions

**conducive to:** likely to produce; helping （an especially desirable result） 有助于（增进；导致）

**take up:** to absorb or use up 吸收

**break down:** to make a substance separate into parts or change into a different form in a chemical process （使）分解

**cope with:** to manage to deal with someone or something 应对；处理

**catch up:** to reach the same level or standard as somebody who was better or more advanced 赶上；追上

**available to:** that you can get, buy or find 可获得的

**carry out:** to do and complete a task 实施，实现

## Comprehension Check

① What is the nitrogen cycle?

② How is the nitrogen cycle in the aquarium?

③ What are the three stages of the nitrogen cycle?

④ When is nitrite produced?

⑤ Once the nitrite spike is reached, what will happen?

## Supplementary Reading

**TEXT A**

微课

### This Is Why Water Is Essential for Life on Earth...and Perhaps the Rest of the Universe

Tom Ireland

1  Water, and specifically liquid water, is deemed so important to the creation and sustenance of life that few scientists entertain the possibility of life existing on worlds without it. The search for extra-terrestrial life by organizations like NASA often boils down to one simple strategy: "follow the water".

2  So why is water so important? Well, there are several reasons, but they all hinge on water's unique chemical properties. The chemical known as H20 is a simple molecule composed of two small positively charged hydrogen atoms and one large, negatively charged oxygen atom. This gives each molecule, and the substance itself, what is called "polarity".

3  The opposing charges mean the different bits of nearby water molecules stick to each other, but it also means that water interacts with the charged elements of other molecules, often helping to break them apart and dissolve them.

4  For the thousands of chemical reactions going on in our cells to happen quickly and efficiently, the molecules involved need to be able to mix freely—they need to be dissolved in something. Water is so good at dissolving substances that it is known as the "universal solvent".

While other substances have similar dissolving power to water, they do not have its chemical stability and its ability to neutralize strong acids and bases.

5  The polarity of water also helps the formation of the delicate membranes that encapsulate all living cells. In water, special fats called lipids line up with their water-loving ends facing outwards and water-hating ends facing inwards, forming a continuous but flexible two-layer film, like the outside of a soap bubble. These lipid membranes play a fundamental role in keeping the complexity of life concentrated in one place, creating individual entities separate from each other and their environment.

6  The polarity of water also helps water molecules to stick to each other, giving water another useful property called cohesion. This means water will get drawn through very thin tubes, even against the pull of gravity, enabling water to flow hundreds of feet from the ground to the tops of tall trees. Handily, H20 also helps transport nutrients, clean away waste, and provides pressure for structural support.

7  Anything else? Ah yes: photosynthesis, the process in plants that creates sugars from sunlight, and which creates the food that feeds the planet's entire food chain, requires—you guessed it—water.

8  In fact, there are so many reasons why water is crucial to life that entire books have been dedicated to it.

9  Could there be strange forms of life out there that don't require water? Of course—it is possible that there is life so different to ours on Earth that we can't even imagine how it works. But it seems highly likely that if there is life out there, it will need a solvent to lubricate the chemical processes that create energy, movement and replication. And we know of no other substance that does that half as well as water.

10  Water's importance as a supportive base substance led the Nobel Prize-winning biologist Albert von Szent-Györgyi to describe it as life's "matrix, mother and medium", while the science writer and physicist Philip Ball wrote this of biology: "You could be forgiven for concluding that the subject is all about proteins and genes, embodied in DNA. But this is only a form of shorthand; for biology is really all about the interactions of such molecules in and with water."

TEXT B

## Seize the Moment

Shi Jianbin

(With less than seven years to go for achieving the UN Sustainable Development Goals, we are just in time for wetland restoration.)

1  From purifying water and providing clean water, to protecting people from storms and floods, maintaining critical ecosystems and biodiversity, and storing large amounts of carbon and mitigating climate change, wetlands are critical to human survival and prosperity.

2  However, globally, wetlands are under threat from both anthropogenic and natural causes. According to the Convention on Wetlands Secretariat, over 35 percent of the world's natural wetlands have been lost in the past 50 years, a rate that is three times that of forests. This means the key ecosystem services that wetlands provide to human beings are in grave peril. It is now critical to raise international awareness about wetlands and to encourage actions to protect and restore these important ecosystems and the services they provide.

3  Feb. 2 marks the 27th World Wetlands Day, the theme being "Wetland Restoration". This emphasizes the urgent need to prioritize the restoration of lost or degraded wetlands and calls for everyone's participation.

4  The theme reflects efforts to heal the planet, promoted by the UN Decade of Ecosystem Restoration (2021-2030) initiative, which is leading and inspiring the restoration of critical ecosystems including wetlands around the world. It provides an excellent opportunity to restore critical ecosystems such as wetlands. The theme is also in line with the ethos of the Wuhan Declaration adopted in November 2022 at the 14th Conference of the Contracting Parties to the Ramsar Convention on Wetlands (COP 14). The declaration reaffirms and emphasizes the importance of protecting, restoring and rationally using wetland resources to address climate change, enhance biodiversity conservation and promote sustainable development, and calls for "urgent measures to achieve the goal of halting and reversing the loss of wetlands globally".

5  Wetland restoration can bring a range of benefits, such as enhanced biodiversity and ecosystem stability, increased carbon storage and reduced emissions for climate change mitigation and adaptation, better protection of people and their property against natural disasters such as floods and storms, and improved tourism development which can secure more sustainable livelihoods for more local people and contribute to poverty alleviation.

6  Thus, wetlands restoration can directly contribute to achieving the relevant goals of the Convention on Wetlands, the Convention on Biological Diversity and the United Nations Framework Convention on Climate Change. While continuing and strengthening traditional in situ approaches for nature conservation, the Post-2020 Global Biodiversity Framework adopted

at CBD COP 15 in December 2022 highlights the important role of ecosystem restoration in reversing biodiversity loss, as reflected in its Target 2: "Ensure that by 2030 at least 30 percent of areas of degraded terrestrial, inland water, and coastal and marine ecosystems are under effective restoration, in order to enhance biodiversity and ecosystem functions and services, ecological integrity and connectivity". The nature-based climate change adaptation and mitigation actions of the UNFCCC are directly related to the restoration and protection of critical ecosystems, including wetlands. Both biodiversity conservation and the response to climate change are closely related to wetland protection and restoration, and are integral parts of the efforts to achieve the UN Sustainable Development Goals.

7  Over the past decade, the Chinese government has introduced and implemented a series of overall plans related to wetland protection and restoration, such as the Programme for a Wetland Protection and Restoration System, the National Master Plan of Major Projects for Protection and Restoration of Key Ecosystems (2021-2035), and the National Wetland Protection Plan (2022-2030), which all have specific contents on and requirements for wetland restoration. The National Wetland Protection Plan, which was released in December 2022, particularly proposes to adopt near-natural measures for the comprehensive improvement and systematic restoration of wetlands in those areas where the ecological functions of wetlands are seriously degraded, such as the lower-reaches of the Yellow River and coastal areas of the Yellow Sea and Bohai Sea. In addition, some tailor-made special action plans on wetland restoration have been developed and put in place, such as the Special Action Plan for Mangrove Protection and Restoration (2020-2025) and the Special Action Plan for Spartina alterniflora Control (2022-2025). The former special action plan requires not only the afforestation of 9,050 hectares of new mangrove forests, but also the restoration of 9,750 hectares of degraded mangroves by the end of 2025, while the latter requires a 90 percent removal rate of Spartina alterniflora nationwide by 2025 to effectively curb the spread of this invasive species.

8  China has come a long way in wetland restoration. Over the past 20 years, the country has increased its mangrove area—from 22,000 hectares to 27,100 hectares today, making it one of the few countries in the world to have increased its mangrove area. In the 13th Five-year Plan (2016-2020), China restored 1,500 kilometers of its coastline and 30,000 hectares of coastal wetlands.

9  The Paulson Institute has not only witnessed but also participated in the restoration of wetlands in China over the past decade, especially in the areas of Spartina alterniflora control

and studies on mangrove protection and restoration strategies. Since 2014, the institute has drawn the attention of the relevant Chinese government agencies to the threats and control of Spartina alterniflora and put forward practical control suggestions.

10  Each individual can take a variety of actions to contribute to wetlands restoration and protection, for example, making conscious choices to minimize our own impacts on wetlands, speaking out persuasively to educate and activate others to get involved in wetland restoration, and acting boldly to engage in local wetland restoration.

11  We are now less than seven years away from the deadline to achieve the UN Sustainable Development Goals and the targets set by the CBD's Post-2020 Global Biodiversity Framework. Reversing the trend of rapid degradation or loss of wetlands is critical to achieving these goals.

### TEXT C

## Improvements Being Made to Nation's Water Quality

Hou Liqiang

1  China made marked progress in controlling water pollution last year, with discharges of major water pollutants continuing to decline, according to the Ministry of Ecology and Environment.

2  China saw 87.9 percent of its surface water rated as being of fairly good quality last year, up by 3 percentage points from 2021, said Huang Xiaozeng, head of the ministry's department of water ecology and environment.

3  The quality of about 0.7 percent of the country's surface water remained below Grade V, the worst quality, down by 0.5 percentage points.

4  Under China's six-tier quality system for surface water, the quality of water is considered as being fairly good if it reaches Grade III and above.

5  The quality of water in the trunk of the Yangtze River, Asia's longest watercourse, has stayed at Grade II for three straight years, Huang continued.

6  In the entire Yangtze basin, he said, 98.1 percent of national monitoring stations reported fairly good water quality last year, up by 1 percentage point year-on-year.

7  The achievements occurred thanks to a series of actions the country has taken in the basin.

8    The ministry has located over 60,000 sewage outlets that discharge into the Yangtze trunk, nine of its major tributaries and Taihu Lake, he said. To date, it has found the sources of sewage in 90 percent of the outlets and addressed violations in more than 20,000 of them.

9    Measures were also rolled out to ramp up the treatment of wastewater in industrial parks, as well as to enhance the management of tailings ponds in mines and the phosphorus sector, he said.

10    Rectification, for example, has been completed in 843 of the total 1,137 tailings ponds found with violations in the Yangtze basin. "We will keep a close eye on the rest of the ponds with violations to ensure that all of them will be adequately rectified," he said.

11    Despite the progress, he said there are still some lingering problems in the Yangtze basin.

12    Due to the lack of a sound cross-department coordination mechanism, for instance, little synergy has been created as different government bodies strive to beef up Yangtze conservation, he noted.

13    Still troubled by the bloom of blue-green algae, he said, some key lakes in the Yangtze basin are yet to see the natural balance of their water ecosystems restored.

14    Huang has also highlighted the ministry's efforts to address black and odorous water bodies as an environmental problem on people's doorsteps that affect their vital interests.

15    In 2022, 40 percent of such water bodies in county-level cities across the country were improved, he said, without disclosing the total number of such water bodies in these cities.

16    He said the ministry will resort to satellite remote sensing and unannounced visits to enhance its supervision over the treatment work of local governments.

17    For those cities with poor performances in collecting wastewater for disposal or direct wastewater discharge, the ministry will notify them with warning letters and also ask local provincial authorities to better play their roles in guiding them in the treatment work.

18    "We will also enhance information disclosure. We established a good mechanism to make public information concerning the treatment of black and odorous water bodies. We will further improve the mechanism to facilitate public supervision," he said.

## Practical English for Ecological Environment

### ● Writing

Write a suggestion letter to the editor of a newspaper about what government officials, society and people can do to conserve water resources and improve water quality.

### ● Unit Project

Work in groups. Focus on one of the following cycles—water, carbon and nitrogen—and talk about its importance to our lives and environment with specific examples.

# Unit 6

# Xi Jinping Thought on Ecological Civilization

Since the 18th National Congress of the Communist Party of China, under the scientific guidance of Xi Jinping Thought on Ecological Civilization, our ecological and environmental protection efforts have seen sweeping, historic, and trans-formative changes. Thanks to its scientific guidance, strict logic and profound visions, Xi Jinping Thought on Ecological Civilization has enriched the Marxist theory of the relationship between man and nature and transformed China's traditional ecological culture. Therefore, it has provided scientific guidance for the understanding of the relationship between man and nature and for the building of the knowledge system of China's ecological civilization. This unit will be unfolded from six aspects:

• Man and nature should coexist in harmony;

• Lucid waters and lush mountains are invaluable assets;

• No welfare is more universally beneficial than a sound ecological environment;

• The country's mountains, rivers, forests, farmlands , lakes and grass form a community of shared life;

• The strictest regulations and laws must be applied in protecting the environment;

• Global ecological conservation requires the joint efforts of all.

At the Eighth National Environmental Protection Conference, these six aspects have been regarded as the six guiding principles that must be followed in advancing ecological civilization. Such move has shed new light on our efforts to battle against pollution and strengthen ecological conservation.

## Warm-up

You are going to watch a video about Taoism and its essence. Watch it twice and complete the following two tasks. Compare your answers with your partner's.

**Task 1**  Identify the main idea.

The video is mainly about the essence and influence of _____.

**Task 2**  Watch the video clip again and fill in the blanks.

(1) It's an _____ that reflects a deep-rooted Chinese worldview.

(2) "The Dao, or the Way, is the approach in accord with _____. Although different countries and various regions have their own cultures, if we pursue _____, there will be no friction and no violence," said Zhong.

(3) Taoism is China's indigenous religion. It's also a religion of _____, as evident in its best-known symbol, the circle of yin and yang.

(4) This philosophy has _____ people in China, as well as those in many Asian countries and more recently in the West, over the course of many centuries. Taoism has become _____.

(5) "The culture of Taoism is universal _____. It has boosted cultural exchanges and mutual learning. The reason why Western people like Taoism so much is mainly because _____," Meng Zhiling added.

**Task 3**  Work in pairs. Discuss with your partners the following questions.

(1) What's the association between traditional Chinese ecological thoughts and Xi Jinping Thought on Ecological Civilization?

(2) What are the influences of Chinese traditional culture on the world?

# Reading Comprehension

**TEXT A**

微课

## Can-Do Spirit

Erik Solheim

(Newly formed Desert Ecological Technology **Alliance** takes its inspiration from the greening of the Kubuqi Desert in Inner Mongolia.)

1  The greening of the Kubuqi Desert in the Inner Mongolia autonomous region is among the greatest environmental achievements of our times. Through hard and smart work, the people of Kubuqi transformed the "Sea of Death" into a **lush oasis**. Now that "Kubuqi spirit" will be guiding a new international organization, the International Desert Ecological Technology Alliance, which was formed this month (Feb., 2023) in Beijing.

2  A new **whopping** investment of 80 billion yuan ($11 billion) is now planned in green energy in Kubuqi. This will be the biggest investment in renewable energies ever in a desert area.

3  The alliance is a partnership between Elion Resources Group, the Secretariat of the UN Convention to Combat Desertification, China State Power and 40 other founding member organizations, research institutions and companies.

4  It will help share technologies, ideas, expertise and investments between dry areas in China, Central Asia, the Middle East, Africa and beyond. A yearly forum will be hosted in Kubuqi. Demonstration bases are to be <u>set up</u> in 10 countries on different continents.

5  The new desert alliance will be rooted in "the Kubuqi spirit".

6  The Kubuqi Desert <u>is situated in</u> the Ordos prefecture of Inner Mongolia. In the mid-Qing Dynasty (1644-1911), slash-and-burn farming resulted in the disappearance of vegetation cover and in the desertification of pastures and grasslands. The area became known as the "Sea of Death". Up until two decades ago, the Kubuqi Desert, the seventh-largest in China, contributed

significantly to the sandstorms in Beijing and Hebei.

7   Yet the first time I visited Kubuqi, I <u>was struck by</u> such a green oasis full of life and greenery in the midst of extremely dry surroundings. It was moving to see how the efforts of thousands of Chinese have been able to turn what was once a very dry desert into an enormously **appealing** place abundant with different forms of life.

8   I have visited many deserts in Africa, Latin America and the Middle East. I am not aware of any success story as impressive as Kubuqi. For sure, people on different continents have been able to green deserts, but the size and the **grit** of Kubuqi still <u>stand out</u>.

9   What was the secret to its success?

10   When I spoke to people, it turned out there was no secret. It was just hard and smart work. It was the **perseverance** of people who put in years of continuous efforts to achieve this great result. It was the innovative spirit of the Elion Group pioneering new approaches, <u>relying upon</u> public-private partnerships involving the local residents.

11   Wang Wenbiao was the key figure in giving this barren land a green soul. Feeling a strong emotional bond to the desert where he was born, he proved one thing: Take good care of Mother Earth, and she will take good care of you and your people.

12   In 1988, 29-year-old Wang took over a saltworks in his home district. The top challenge facing the factory was an expanse of sand driven by endless sandstorms that kept <u>creeping in</u> around the salt lake where production was based. He and his fellow men worked **laboriously** and **painstakingly** to try out all types of methods to improve life and production, including building a road, taking a portion of earnings to finance sand control and assigning a significant portion of the factory's workforce to plant trees.

13   They didn't get everything right from the beginning. But they learned from their mistakes and gained experience. The once-young idealist Wang became a major business leader known for his **charismatic** ability to inspire people.

14   Early on they understood that only a private company could help generate commercial value from the desert. It was <u>in line with</u> the opening-up policy of China from the 1980s onwards. It was based on a very clear understanding that over time, you cannot bring people out of poverty through handouts, or transform a harsh landscape just by being **altruistic**. You need an approach that creates value to make people rich while taking care of the desert. His company Elion became a sand-control pioneer.

**15**   In 2017, I had the **privilege** to honor Wang with the United Nation's highest prize for the environment, the UN Champions of the Earth. It was an award for a lifetime of leadership in green industries.

**16**   At the core of the "Kubuqi spirit" is the idea that a desert is not a threat but an opportunity for economic growth and poverty alleviation. The essence of the "Kubuqi spirit" is to **merge** the ecology and economy into the win-win proposition that you create value and jobs from the desert, and you do it by going green.

**17**   The Kubuqi miracle has been linked to three sources of income for people in the desert.

**18**   The first **originates** in the enormous space in the desert, which is suitable for renewable energy. It has been used in Kubuqi to host huge solar farms. It's beautiful to see dry landscapes used in this way to create energy which can <u>bring up</u> more water for agriculture, and which can power homes, and electric cars and buses.

**19**   Now a massive new phase of renewable energy in Kubuqi has been **initiated** in a partnership between Elion, Three Gorges Group and others. Around 80 billion yuan ($11.86 billion) is to be invested in a program of 8 gigawatt solar, 4 gigawatt wind and 4 gigawatt upgraded coal, along with energy storage, to adjust for peak demands. This will provide a massive injection of green electricity to Beijing, Tianjin and Hebei Province.

**20**   The second is to grow products which can survive and thrive in dry places. Elion's liquorice plantations are known as a "sweet deal" for the company, the growers and the environment. Elion gets a quality cash crop. Local households benefit from technical expertise and guaranteed sales. The environment benefits from the natural nitrogen fixation and soil improvement properties of the plants.

**21**   Tourism is a third source of income. Tourism is probably now the biggest job creator in the world. Kubuqi is a beautiful place not very far from population centers in Inner Mongolia and neighboring provinces. I took my teenage son to Kubuqi, and he was **astonished** by all the leisure opportunities there—to go camping in the desert, riding on camels and horses, and, for a teenager perhaps most temptingly, to ride race cars in the desert. There are beautiful hotels which serve great Chinese and Mongolian cuisines, for instance, traditional Mongolian lamb with vegetables from the desert.

**22**   From plantations and tourism to solar energy, the benefits of a whole range of activities have already restored more than one-third of the Kubuqi Desert and lifted over 100,000 farmers out of poverty.

**23**   These huge achievements in Inner Mongolia are exactly what President Xi Jinping <u>seeks to</u> **encapsulate** when he says that lucid waters and lush mountains are invaluable assets. Or green is gold.

**24** The US space agency NASA recently published a report saying that, contrary to what many people think, the surface of our small planet is greener than it was in the past. And that is largely thanks to tree planting in China, in Kubuqi, and also in the Xinjiang Uygur autonomous region, Gansu Province, Hebei Province and other parts of the country. This is a great service to China and to humanity.

**25** The Desert Ecological Technology Alliance is an exciting initiative, which will provide a platform for the world to be inspired by the Kubuqi spirit and to exchange views and learning between desert areas in all corners of the world.

## New Words

**alliance /əˈlaɪəns/ n.**
(with, between) a close agreement or connection made between two countries, groups, families 结盟，联盟

**lush /lʌʃ/ adj.**
produced or growing in extreme abundance 苍翠繁茂的；茂盛的；丰富的

**oasis /oʊˈeɪsɪs/ n.**
a fertile tract in a desert (where the water table approaches the surface) 绿洲；宜人之地

**whopping /ˈwɑːpɪŋ/ adj.**
(used informally) very large 巨大的；非常大的；异常的

**appealing /əˈpiːlɪŋ /adj.**
able to attract interest or draw favorable attention 引起兴趣的；动人的；恳求的

**grit /ɡrɪt/ n.**
a hard coarse-grained siliceous sandstone 沙砾；粗沙石；勇气；决心

**perseverance /ˌpɜːrsəˈvɪrəns / n.**
determination to keep to trying to do sth in spite of difficulties 不屈不挠；毅力；坚韧不拔

**laboriously /ləˈbɔːriəsli / adv.**
in a laborious manner 艰难地；辛勤地

**painstakingly /ˈpeɪnzteɪkɪŋli / adv.**
in a fastidious and painstaking manner 费力地；苦心地

**charismatic /ˌkærɪzˈmætɪk / adj.**
possessing an extraordinary ability to attract 有魅力的

**altruistic /ˌæltruˈɪstɪk / adj.**
showing unselfish concern for the welfare of others 利他主义的；无私的

**privilege /ˈprɪvəlɪdʒ/ n.**
sth enjoyable that you are honored to have the chance to do 特权；荣幸；特别恩典

**merge /mɜːrdʒ/** *v.*

(cause two things to) come together and combine; fade or change gradually (into sth else) 合并

**originate /əˈrɪdʒɪneɪt/** *v.*

have as a cause or beginning 发起；开始；起源于

**initiate /ɪˈnɪʃieɪt/** *v.*

set (a scheme, etc.) working 开始，着手

**astonish /əˈstɑːnɪʃ/** *v.*

(cause to) produce great surprise or wonder in sb 使惊讶，使大为吃惊

**encapsulate /ɪnˈkæpsjuleɪt/** *v.*

put in a short or concise form; reduce in volume 压缩；概括；装入胶囊

## Phrases and Expressions

**set up:** If you set up somewhere or set yourself up somewhere, you establish yourself in a new business or new area. 开办；开创事业

**be situated in:** to exist or have been built in a particular place or position 坐落在, 位于

**be struck by:** to be very impressed by or pleased with (something or someone) 被……所迷惑

**stand out:** to be much better or more important than somebody/something 出色；杰出

**rely upon:** put trust in with confidence 依靠；依赖；取决于

**creep in:** enter surreptitiously 悄悄混进

**in line with:** in agreement with 按照；与……一致

**bring up:** 提出；教育；养育

**seek to:** 追求，争取，设法

## Exercises

**I Understanding the Text**

1. Recognize the sequence in which the passage presents its most important information on the Kubuqi spirit. Fill in the form with the correct choice from the following list.

① The secret to Kubuqi success.

② "The Kubuqi spirit" will be guiding a new international organization.

③ The new desert alliance will be rooted in "the Kubuqi spirit".

④ The Kubuqi miracle has been linked to three sources of income for people in the desert.

⑤ The achievements and significance of the greening of the Kubuqi Desert.

| Parts | Paragraphs | Main ideas |
| --- | --- | --- |
| Part One | Paras.1-4 | |
| Part Two | Paras.5-8 | |
| Part Three | Paras.9-16 | |
| Part Four | Paras.17-22 | |
| Part Five | Paras.22-25 | |

2. Focus on the parallel sentences in the text. Pay attention to the sentence patterns and fill in key information.

① It will help share _____ between dry areas in China, Central Asia, the Middle East, Africa and beyond.

② It was just _____. It was _____ who put in years of continuous efforts to achieve this great result. It was _____ of the Elion Group pioneering new approaches, relying upon public-private partnerships involving the local residents.

③ Local households benefit from _____. The environment benefits from _____.

④ The first _____, which is suitable for renewable energy. The second is to _____. _____ is a third source of income.

**II Focusing on Language in Context**

1. Key Words & Expressions

A. Fill in the blanks with the words given below. Change the form where necessary. Each word can be used only once.

**alliance oasis appealing perseverance laboriously privilege merge originate astonish**

① It is a great _____ to know you.
② This desert of unattractiveness has its _____.
③ I was _____ at the news of his escape.
④ The two countries entered into a defensive _____ with each other.
⑤ We must have _____ and confidence in ourselves.
⑥ His department will _____ with mine.
⑦ Its delicate aroma is sweet and _____.
⑧ He was trudging _____ on the rugged mountain path.
⑨ Who _____ the concept of stereo sound?

B. Fill in the blanks with the phrases given below. Change the form where necessary.

**set up; creep in; seek to; be situated in; be struck by; rely upon; bring up; in line with; stand out**

① He _____ the crowd, and was lost to sight.
② The bill _____ repeal the existing legislation.

## Unit 6  Xi Jinping Thought on Ecological Civilization

③ A fund will be _____ for the dead men's families.

④ The one who claims to be smart _____ easily _____ a simple trap.

⑤ Our hotel _____ the middle of the city.

⑥ The enhanced rail and road transportation networks between China and Kazakhstan will _____ a win-win situation as China has huge demand for imports from the Central Asian country.

⑦ A man should _____ his own power.

⑧ My plan is _____ your suggestions.

⑨ His height makes him _____ in the crowd.

2. Usage

A. Complete the following sentences in the text by using noun clause led by "what".

① It was moving to see how the efforts of thousands of Chinese have been able to turn _____ abundant with different forms of life.（把曾经非常干燥的沙漠变成一个非常吸引人的地方）

② These huge achievements in Inner Mongolia are exactly _____ _____ when he says that lucid waters and lush mountains are invaluable assets.（习近平主席力图概括的事物）

③ The US space agency NASA recently published a report saying that, contrary to _____ _____, the surface of our small planet is greener than it was in the past.（很多人的认知）

B. Complete the following sentences in the text by using appositive clause.

① You need an approach _____ while taking care of the desert.（它创造价值，让人们变得富有）

② It was based on a very clear understanding _____ _____. （即随着时间的推移，你无法通过施舍让人们摆脱贫困，或仅仅通过利他主义来改变恶劣的环境）

③ At the core of the "Kubuqi spirit" is the idea _____ _____. （即沙漠不是威胁，而是经济增长和扶贫的机会）

④ The essence of the "Kubuqi spirit" is to merge the ecology and economy into the win-win proposition _____.
（即你要从沙漠中创造价值和就业机会就得走绿色发展道路）

### III Translation

Translate the following passage into English.

过去 10 年，海平面升高和森林砍伐的速度都是前所未有的。生态恶化、物种灭绝、臭氧层被破坏、温室效应、酸雨等一系列环境问题已经严重影响到人类的生存环境。环境恶化造成的问题之一就是缺水。目前全世界 40%以上的人口，即 20 多亿人，面临缺水问题。据预测，未来 25 年全球人口将由 60 亿增长到 80 亿，环境保护面临更大的压力。中国作为一个发展中国家，面临着发展经济和保护环境的双重任务。从国情出发，中国在全面推进现代化的进程中，将环境保护视为一项基本国策。众所周知，对生态环境和生物多样性的保护是环保工作的重点。我国野生动植物物种丰富，仅脊椎动物就有6000多种，高等植物3万多种。

**IV Pair Work**

Talk about the great achievements China has made in ecological construction. Illustrate your ideas with specific examples.

**TEXT B**

### From 50 to 50

Zhang Jianyu

(Aligning a beautiful China with global partners can help **avert** climate catastrophe.)

1　Exactly 50 years ago in Stockholm, when the Swedish government received a Chinese delegation participating in the first United Nations Conference on the Human Environment, neither side realized the significance of this event would go far beyond that it was China's first official UN-sponsored activity following its historic return to the UN family in 1971.

2　China immediately embraced the still **nascent** concept of environmental protection and inaugurated a half-century sustainability journey that no one could have **envisioned** back then. Although a latecomer that is still going through its uniquely fast and condensed industrialization process, China has demonstrated remarkable willingness not only to learn, but also to cooperate with the international community on finding solutions to all daunting environmental threats.

3　Actually, China was so open and creative that in 1992 it established a dedicated cooperation mechanism, China Council for International Cooperation on Environment and Development. The CCICED <u>is composed of</u> experts working on almost all related topics in the field, 50 percent of them are Chinese experts and 50 percent foreign. Two days ago, the CCICED celebrated its 30th anniversary, which in itself **testifies** to the progress that China has made. One everlasting theme that the

CCICED has witnessed is that while the green agenda has steadfastly become the central theme of development, China continues to make progress in economic and social development and poverty alleviation, demonstrating that social, economic and environmental progress can go hand-in-hand.

4  The year 2050 has emerged as the critical milestone year for the global community beyond the 2030 target year for realizing the 17 Sustainable Development Goals of the UN. The year 2050 is significant, because it is an international scientific consensus that, in order to prevent the worst climate damages, global net human-caused emissions of carbon dioxide need to reach net zero around 2050. By that year, signatories of the Paris Agreement are required to plan their mid-century carbon control strategies.

5  <u>Today, 132 countries have pledged to achieve carbon neutrality by 2050 and many others are joining them on their way</u>. Long-term strategies have a critical role to play in this transition because only through setting up both bold and collective goals and then back-casting and **extrapolating** from the long-term goal can humanity achieve the necessary balance between development and natural resources.

6  Coincidentally, 2049 is the 100th anniversary of the founding of the People's Republic of China, which is the year set for the country to have achieved its goal of becoming a modern prosperous country with a harmonious relationship between man and nature. <u>One of the most important aspects of this goal is building an ecological civilization, the concept that China has been **pioneering** and has included in its Constitution.</u> With a strong central planning mechanism and a track record of achieving its environmental targets during its five-year plans, China has the unique experience and capabilities to achieve this necessary balance domestically, demonstrating to the world how important it is to equip ambitions with step-by-step actions.

7  Despite all the turbulence and chaos, the 21st century will undoubtedly be recorded in modern history as the "green century", not only because environmental awareness planted in the last century has now come to **fruition**, but also because the health and well-being of people are more keenly threatened by the "triple planetary crises" of climate change, nature and biodiversity loss, and pollution. The situation has been made even worse in recent years by the devastating COVID-19 pandemic and by geopolitical tensions that are slowing global cooperation. <u>Deep decarbonization goals can only be achieved through the sharing of resources, knowledge and experiences and can only be achieved when a broader constellation of countries, cities and companies are engaged.</u>

8    China has shown itself willing to act to address these triple planetary crises and is a convincing contributor, based not only on what it has learned and practiced in the past 50 years, but also on the green pathway the country has pioneered for itself toward mid-century. Through cooperation with the European Union and the United States, China has established the world's largest carbon emissions trading system, which now covers the power sector, representing more than 40 percent of national carbon dioxide emissions.

9    While much of the world has seen a rise in pollution in recent years, global pollution has decreased due entirely to China's impact since 2013. Without China's significant decline in pollution, global average pollution would have increased in that time. Through hosting the COP 15, China has strengthened its entire ecological governance and it has now pledged to plant and cultivate 70 billion trees by 2030.

10    With China's influence increasingly felt in the global arena, China does not hesitate to supply green goods to the global society and shoulder its differentiated responsibilities. With the launch of the Belt and Road Initiative International Green Development Coalition, China is applying what it has learned from the CCICED through the BRIGC to the countries that it has policy, infrastructure, trade, financial and cultural relationships with and together they  are exploring new ways for China's learnings and experiences to spur and accelerate coupled climate and economic progress in partner countries. One **groundbreaking** effort of this is China's announcement of stopping building coal-fired power plants in Belt and Road countries and endeavoring to help other countries build green and low-carbon energy systems.

11    As a good participant and contributor to the global sustainability agenda, it is now the time for China to be more active and become a torchbearer in the global green agenda. With a production capacity of 80 percent of global solar PVs and 70 percent of the global battery production, China should no doubt uphold the global drive toward green and low carbon energy. What the world does this decade will dictate whether the global goal of limiting global warming to 1.5 C—and thus avoid the worst impacts—is within reach. The world needs China to take a lead role in delivering on that promise. One way China can do that is through a modernization vision for a "Beautiful China" by 2035 that links ambitious climate targets with economic and social development goals, while aligning global partners on comparable targets within the same time frame.

## New Words

**avert** /əˈvɜːrt/ *v.*
prevent sth bad or dangerous from happening 防止；避免
**nascent** /ˈnæsnt/ *adj.*
being born or beginning 新生的；尚不成熟的
**envision** /ɪnˈvɪʒn/ *v.*
imagine; conceive of; see in one's mind 想象；设想
**testify** /ˈtestɪfaɪ/ *v.*
serve as evidence of; bear witness to; give evidence of 为……作证；证明
**extrapolate** /ɪkˈstræpəleɪt/ *v.*
draw from specific cases for more general cases 推测；推知
**pioneer** /ˌpaɪəˈnɪr/
*v.*
begin or help in the early development 开辟；首倡
*n.*
先锋；开拓者；先驱
**fruition** /fruˈɪʃn/ *n.*
something that is made real or concrete 取得成果；成就；实现；完成
**groundbreaking** /ˈɡraʊndbreɪkɪŋ/ *adj.*
being or producing something like nothing done or experienced or created before 创新的

## Phrases and Expressions

**be composed of:** to be made of a particular substance or substances 由……组成
**pledge to:** to formally promise to give or do something 立誓
**endeavor to:** to try to do something 争做

## Exercises

### I Comprehension Check

① What is the CCICED composed of?

② Why is the year 2050 significant?

③ Why do long-term strategies have a critical role?

④ What is the "triple planetary crises"?

⑤ What is BRIGC?

___

**II Translation**

Translate into Chinese the underlined sentences in the article.

① Today, 132 countries have pledged to achieve carbon neutrality by 2050 and many others are joining them on their way.

___

② One of the most important aspects of this goal is building an ecological civilization, the concept that China has been pioneering and has included in its Constitution.

___

③ Deep decarbonization goals can only be achieved through the sharing of resources, knowledge and experiences and can only be achieved when a broader constellation of countries, cities and companies are engaged.

___

④ With China's influence increasingly felt in the global arena, China does not hesitate to supply green goods to the global society and shoulder its differentiated responsibilities.

___

⑤ As a good participant and contributor to the global sustainability agenda, it is now the time for China to be more active and become a torchbearer in the global green agenda.

___

**III Group Work**

China has been closely cooperating with other countries in ecological construction. Illustrate it with a specific example.

**TEXT C**

微课

## Finding Ways to Keep the Water Flowing (1)

(Ahead of the UN 2023 Water Conference, a three-day gathering that will start on Wednesday,

World Water Day, Li Guoying, minister of Water Resources, talks about China's endeavors to enhance water management and promote international cooperation in an exclusive interview with *China Daily* reporter Hou Liqiang. Here are the excerpts.)

**How would you comment on the achievements China has made in water resource management over the past 10 years?**

1   China has made remarkable achievements in water resource management over the past 10 years.

2   <u>First, the national capability to conserve water resources and use them efficiently has continued to improve.</u> Despite an average annual economic growth rate of about 6 percent, China's annual water consumption has been maintained below 610 billion cubic meters.

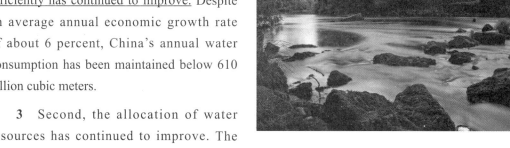

3   Second, the allocation of water resources has continued to improve. The annual water supply capacity nationwide is <u>in excess of</u> 890 billion cubic meters.

4   Thirdly, the capacity for the environmental protection and governance of rivers and lakes has improved. We have implemented the River Chief System (a network of leading officials responsible for overseeing the conditions of specific bodies of water), launched the "Mother River" restoration program and controlled the excessive exploitation of groundwater. More and more water bodies and river basins have been **rejuvenated**. Many rivers that were previously dry for extended periods of time once again flow.

5   Fourthly, as China strives to provide household water security to rural areas, 87 percent of the rural population now has access to tap water, up 11 percentage points from 2015.

6   These accomplishments have happened because we pursue a people-centered development philosophy and always see public aspirations for a better life as our goal.

7   We also follow natural law, prioritize environmental protection and green development and implement new development concepts accurately and completely. We balance the relationship between development and protection, and endeavor to promote harmony between people and water resources.

8   We follow the rule of law in water governance and reform and innovation, and make a consistent effort to improve water-related laws and regulations, such as the Yangtze River Protection Law.

**How would you evaluate the historical significance of the South-to-North Water Diversion Project? What experiences are worth sharing with the international community?**

9   As the largest cross-river basin water transfer project in the world, construction of the project officially began in December 2002 after half a century of planning and assessment. As of December 2014, the first phases of the East and Middle Routes have become operational. To date, they have transferred 60.5 billion cubic meters of water, directly benefiting more than 150 million people.

10   The project has led to the accumulation of valuable experience in implementing major cross-basin water transfer projects:

• First, the importance of adhering to the national plan, with local interests subordinate to the overall situation.

• Second, the importance of concentrating resources to accomplish great things by promoting the project at the central level and concentrating on securing elements such as funding and land use, as well as coordinating resettlement of those displaced by construction.

• Third, the importance of respecting law, scientifically and **prudently** demonstrating proposals, and emphasizing environmental protection. In addition, the importance of emphasizing both willpower and a balance between humans and water resources.

• Fourth, the importance of taking a holistic approach to planning, taking into account the conditions in the four major river basins of the Yangtze, Huaihe, Yellow and Haihe rivers, as well as the demands of regions and industries.

• Fifth, the importance of focusing on both water conservation and pollution control.

• And sixth, the importance of accurately and precisely regulating water transfers, hammering out detailed water allocation plans and accurately **dispatching** water from the source to users.

11   In your opinion, what are some of the experiences the international community can learn from? Can you share a few examples that have impressed you?

12   China faces one of the most challenging water management situations in the world. It has a large population but limited water resources. The **spatiotemporal** distribution of those resources is uneven. There is a mismatch between water distribution and regional socioeconomic factors. The carrying capacity of the water environment is limited.

13   China has accumulated some experience in improving water resource management at home while conducting international cooperation abroad.

**14** Since the 1970s, the long-term, large-scale exploitation of groundwater resources has led to the serious overexploitation of groundwater in North China, as well as ecological and environmental problems, including a decline in groundwater levels and the shrinking of river and lake surfaces.

**15** In 2019, with State Council approval, the Ministry of Water Resources and relevant departments issued the Action Plan for Comprehensive Treatment of Over-exploitation of Groundwater in North China. There has been an overall rise in groundwater levels in North China, with shallow groundwater and deep confined water in the treatment area rising by an average of 2.25 meters and 6.72 meters, respectively, compared to 2018.

**16** We have implemented the River Chief system. At present, there are 1.2 million river chiefs at the provincial, city, county, township and village levels nationwide. Under their management, targeted measures have been taken to address problems in each river and lake. As a result, the appearance of rivers and lakes has undergone historic change.

**17** We have contributed water management solutions to global water governance. On April 27, 2019, the International Standardization of Small Hydropower was included in the list of outcomes of the Second Belt and Road Forum for International Cooperation as an important achievement of BRI international cooperation. The China-headquartered International Network on Small Hydro Power worked together with the United Nations Industrial Development Organization to publish the Technical Guidelines for Small Hydropower, which is the first systematic international standard for the small hydropower industry in the world.

**18** Since the launch of the Lancang-Mekong Cooperation Mechanism in 2016, the water resource authorities of the six countries in the basin have implemented the consensus reached by their leaders. Through consistent efforts to strengthen water resource cooperation, they have achieved fruitful results. To help countries in the basin better cope with climate change, flooding and droughts, the ministry has been providing year-round **hydrological** data to the Mekong River Commission since November 2020. Before that, China provided flood season hydrological data to the commission for 18 **consecutive** years.

**19** The ministry is accelerating the construction of digital twins for river basin and water resource management projects. What is the current progress of these projects?

**20** Digital technology is increasingly becoming a driving force in innovative development. Based on the demand for the development of water resource governance, the ministry has proposed accelerating the creation of digital twin projects, and progress has been made in three aspects

**21**　First, the ministry has issued a series of top-level design documents, clarifying the goals, tasks and layout of river basin digital twins. By 2025, key areas along major rivers, lakes and their main tributaries <u>are slated to</u> complete the creation of the digital twins.

**22**　Second, pilot projects are being promoted. The construction of digital twin river basin is a complex and systematic project. The ministry identified 94 priority tasks across the country to begin work on last year.

**23**　Third, the ministry has endeavored to support water management businesses in realizing the four functions of forecasting, early warning, simulation and contingency planning with the help of digital technology.

**24**　The ministry will coordinate the creation of digital twins for river basins, water supply networks and water resource engineering. By the end of the 14th Five-Year Plan period (2021-25), it is slated to create a digital twin system that encompasses the four functions previously mentioned.

### New Words

**rejuvenate** /rɪˈdʒuːvəneɪt/ *v.*
restore (a river or stream) to a condition characteristic of a younger landscape 使（河流、溪流）回春；使焕然一新

**prudently** /ˈpruːdntli/ *adv.*
in a prudent manner 谨慎地；慎重地

**dispatch** /dɪˈspætʃ/ *v.*
send sb/sth off to a destination or for a special purpose 调遣；发送

**spatiotemporal** /ˌspeɪʃioʊˈtempərəl/ *adj.*
of or relating to space and time together (having both spatial extension and temporal duration) 时空的

**hydrological** /ˌhaɪdrəˈlɒdʒɪkəl/ *adj.*
of or pertaining to hydrology 水文学的

**consecutive** /kənˈsekjətɪv/ *adj.*
one after the other 连续的；连贯的

### Phrases and Expressions

**in excess of:** over 超过

**subordinate to:** less important than something else 比……次要

**hammer out:** to produce (something, such as an agreement) by a lot of discussion or

# Unit 6 Xi Jinping Thought on Ecological Civilization

argument 反复敲定；推敲出

**be slated to:** to plan that something will happen at a particular time in the future 计划；安排

### Translation

① First, the national capability to conserve water resources and use them efficiently has continued to improve.

② The project has led to the accumulation of valuable experience in implementing major cross-basin water transfer projects.

③ The spatiotemporal distribution of those resources is uneven.

④ To help countries in the basin better cope with climate change, flooding and droughts, the ministry has been providing year-round hydrological data to the Mekong River Commission since November 2020.

⑤ Based on the demand for the development of water resource governance, the ministry has proposed accelerating the creation of digital twin projects, and progress has been made in three aspects.

## Supplementary Reading

### TEXT A

#### County to Boost Protection Efforts

Zhang Yangfei  Li Yingqing

1  Yuanyang County in Yunnan Province will strengthen the protection of its terraced rice fields, a UNESCO World Heritage site.

2  Li Yi, director of the Yuanyang Management Committee of World Heritage Hani Terraces, said the county will continue to renovate its ancient buildings, which are mostly the homes of people from the Hani

ethnic group, mobilize local people to improve sanitary conditions and repair and upgrade the infrastructure.

3   He said local authorities will also establish a monitoring system and databases to strengthen daily dynamic monitoring of the surrounding forests, villages, terraces and water, as well as agricultural production and natural disasters.

4   They will establish an expert committee and a think tank for the protection of the terraced fields and regularly invite scholars and experts at the national, provincial and prefecture levels to conduct field research and offer advice related to protection.

5   Yuanyang aims to develop the site into an ecological demonstration area. It will reinforce the conservation of forests, restore original buildings and strengthen the integrity of local architecture, and it will award subsidies to farmers who turn dry land into paddy fields.

6   Renovation of the water system will be another major task. The authorities will repair canals, promote a water-saving irrigation system, build a modern drainage network and accelerate construction of reservoirs to improve water utilization.

7   Li said the management committee has strengthened supervision of key forest areas to prevent any form of logging. By the start of this year, 1.7 hectares of forest in the heritage area had been restored, and soil erosion had been treated across 74.67 square kilometers. The forestation level at the site is now more than 70 percent.

8   As the water system plays a critical role in making the terraces possible, the management committee has assigned special teams, dubbed "ditch runners", to supervise and maintain the ditches that flow through villages and into the fields. The teams have also focused on two key terraced areas and maintained 140 ditches that span 451.33 kilometers.

9   Li noted that some villagers still lack a proactive approach to heritage protection. Although local villages have shaken off extreme poverty and have started to develop tourism, the results cannot be consolidated by simply relying on planting on the terraces.

10   To solve the problem while protecting the environment, Yuanyang will continue to improve local services, integrate agriculture with the tourism and cultural industries, and make efforts to build high-quality agricultural brands, he said.

**TEXT B**

## People at the Center: From Poverty Eradication to Rural Revitalization

Xin Ge

1   Chinese President Xi Jinping visited north China's Shanxi Province ahead of the Spring Festival on January 26.

2   President Xi has for 10 consecutive years visited people at the grassroots level, spoken with villagers and frontline workers, and inspected their living conditions prior to the Spring Festival, which is an important occasion of family reunion for the Chinese people.

3   In the inspection tour on Wednesday, President Xi visited villagers' homes and the post-disaster reconstructed facilities in flood-hit areas of Shanxi last October and inspected local work to restore farming, to ensure everyone's access to heating in winter. He called for more efforts to consolidate the country's achievement in poverty eradication and fully advance rural vitalization.

4   As a large agricultural country with a long history, China's rural regions are a backbone of the country's overall development. When the People's Republic of China was founded in 1949, the majority of people lived below the poverty line and the GDP per capita was a mere 119 yuan in 1952, equivalent to about $60 at the time. Since the beginning of the reform and opening-up, policies and frameworks have been made to speed up the agricultural modernization process, and over 700 million rural residents have been lifted out of poverty.

5   Statistics show that 98.99 million impoverished rural residents have shaken off poverty since the 18th National Congress of the Communist Party of China (CPC) in 2012. All 832 impoverished counties and 128,000 impoverished villages have been removed from the poverty

list. China has succeeded in eradicating absolute poverty, meeting the target set out in the UN 2030 Agenda for Sustainable Development 10 years ahead and contributing to over 70 percent of the global poverty reduction.

6  The swift accomplishment of poverty reduction in China could not be achieved without firmly adhering to a "people-centered" philosophy, which is definitely not a rhetorical slogan but a true commitment by the CPC.

7  Poverty eradication has been one of President Xi's key initiatives since 2012. The previous poverty alleviation efforts mainly targeted impoverished geographical units. In 2014, the "targeted poverty alleviation" program was released, making the targeting more precise by identifying poor households and offering a poverty-alleviation plan on a household basis, which is clearly people-centered. As President Xi emphasized, "Without solving the poverty problem in rural areas, China cannot become a moderately prosperous society."

8  In 2018, the CPC Central Committee and the State Council issued the "2018-2022 Strategic Planning for Revitalization of Rural Areas", which pointed out that developing intelligent agriculture and digitalization in agricultural production, operations, management, and services are vital for realizing agricultural modernization. Moreover, the rural living environment should be improved to make villages livable and beautiful. This historic document outlines a roadmap for success in building a moderately prosperous socialist country where there should be zero poverty in rural areas.

9  Over the years, it can be truly observed that China has stepped up efforts to transform its rural governance structures, improving living conditions of villagers, upgrading rural infrastructures, and providing public service to rural residents. According to a recent report from the National Bureau of Statistics, China's retention rate of nine-year compulsory education has hit 95.2 percent and the enrollment ratio for high school education reached 91.2 percent.

10  In February 2021, President Xi made the announcement that China had secured a complete victory in its fight against poverty, but it's not over. The National Rural Revitalization

Bureau was built in the same month, replacing the Poverty Alleviation Office to continue improving the livelihood of rural people.

**11** President Xi has on various occasions stressed that the declaration of eradicating poverty is not the finish line, but one step on the much greater journey of rural revitalization, which includes further developing rural industries, strengthening ecological civilization construction, deepening rural reforms and enhancing and improving rural governance.

**12** The emphasis on the consolidation of the poverty-eradication achievements reflects the Party's clear recognition that despite the elimination of absolute poverty, some people still live above but close to the poverty line. These people remain vulnerable to unforeseen emergencies such as job loss, harvest failure, serious disease in the family or the pandemic and risk slipping back into poverty. Therefore, complete eradication of poverty for them entails further endeavors.

**13** At the local level, the governments not only promote workers training programs at vocational colleges to prepare the rural workforce for better employment, but also offer financial assistance to poor households to develop their products. At the provincial level, a range of paired assistance between eastern and western regions of China was encouraged with an eye on the promotion of coordinated regional development as well as mobilizing the whole society to participate.

**14** At the national level, in June 2021, China issued 51.2 billion yuan ($8.04 billion) worth of 64 rural revitalization bonds in tandem with the national strategy on rural development. The funds raised would be used in areas such as guarantee grain purchase and storage, smart energy facilities in rural areas and rural road construction.

**15** What is important to note here is that China achieved this grand and historic success not only because of its national strength but also due to the strong leadership of the CPC. The Party has not only introduced the vision and direction for rural revitalization but also constantly supervised the progress of the implementation of projects designed from the central to the different levels of local governments.

**16** As one can tell, President Xi has visited many impoverished villages and talked with the frontline administrators during the campaign over the past years. The great efforts in poverty alleviation and rural revitalization vividly interpret the willingness of the CPC to serve the people wholeheartedly.

17    The Party's focus on rural revitalization reflects the importance rural development plays in the pursuit of a "strong, democratic, civilized, harmonious and modern socialist country", which is the country's long-term goal. The rural revitalization program is going to be a great success, because it is carried out again with the "people-centered" philosophy, which has been evidently embodied in the efforts of the CPC's combat against poverty and the COVID-19 pandemic.

**TEXT C**

### Finding Ways to Keep the Water Flowing (2)

**In recent years, China has faced natural disasters resulting from extreme weather conditions. What progress has the ministry made in addressing climate challenges?**

1    China pursues a people and life-first approach as it works to strengthen the construction of its water and drought control systems.

2    At present, a flood control system consisting of reservoirs, watercourses, embankments and flood detention areas is essentially complete along major rivers. Through the measures of interception, diversion, storage, detention and discharge, it has achieved the ability to largely protect itself against the largest floods to have occurred since the founding of the People's Republic of China in 1949.

3    The country is also able to ensure the security of urban and rural water supply and minimize losses to drought.

4    In terms of non-engineering measures, monitoring, forecasting and warning capabilities have been improved. There are now 120,000 monitoring stations for water resources management, and the time required to collect information from them all dropped to just 15 minutes. The flood forecast accuracy for major rivers in the south and north stands at 90 percent and 70 percent, respectively.

5    In recent years, the ministry has successfully overcome severe flooding in major rivers and lakes thanks to its engineering and non-engineering systems.

6    We have also been able to mitigate rare droughts in the Yangtze River and Pearl River

basins, guaranteeing safety of lives and property, and the security of water supply and food production.

**What kind of exchange and collaboration has the Ministry of Water Resources had with UN agencies in terms of water resources management?**

7   The ministry has been collaborating with UN agencies for a long time. Since joining the United Nations Secretary-General's Advisory Board on Water and Sanitation, it has played an active role in promoting the establishment of a separate water goal in the UN 2030 Sustainable Development Agenda. Through an agreement with UNESCO, the ministry established the International Research and Training Center on Erosion and Sedimentation in China. It has also signed a memorandum of understanding on cooperation with the Organization for Economic Cooperation and Development to promote the drafting of international standards for small hydropower. I will give two examples in detail.

8   First, China has been an active participant in UNESCO's International Hydrological Program for a long time. It established the Chinese National Committee for the IHP in 1979, and has been an active partner and important contributor to the program. Over the years, within the program's framework, both parties have cooperated on hydrological research, water education and training, and cultural exchange. A Chinese expert currently serves as chair of the IHP's Intergovernmental Council.

9   Second, China, UNIDO and the UNDP created the International Network on Small Hydro Power. It now has 480 members from 80 countries and regions and has three regional centers in Asia, Africa and South America. The network has rolled out international cooperation on rural energy development, environmental protection and poverty relief. The Lighting up Africa with Small Hydro Power project has been recognized by the UN and welcomed by recipient countries.

10   Looking to the future, we will push forward the Belt and Road Initiative and the Global Development Initiative through multilateral UN platforms, and work with countries and regions worldwide to tackle increasingly severe water challenges.

**This year marks the 10th anniversary of the Belt and Road Initiative. What landmark water resource management projects has China rolled out in signatory countries? What plans are there for the future?**

11   China is not just the founder of the Belt and Road Initiative, but is also a responsible

partner. In recent years, cooperation with BRI signatories has made remarkable contributions to socioeconomic development in participating countries.

**12** The first is that China has helped guarantee livelihoods in BRI countries. In many countries in Asia, Africa and Latin America, China has provided technical consultation for the comprehensive planning of water resources, river basins and flood control. Also, a number of water resource management projects have been implemented to help address insufficient water supplies and agricultural development, and to strengthen the ability to prevent flooding and mitigate droughts. The Lancang-Mekong Sweet Spring Project, for example, has addressed the difficulties of some 7,000 people in Cambodia, Laos and Myanmar in getting safe drinking water.

**13** The second is that China has injected new impetus into the economic development of BRI countries. Chinese planning and design companies have participated in the construction of a number of multifunctional water resource management and hydropower projects. These have not only helped guarantee power supply, but have also promoted industrial upgrading and created many jobs.

**14** Thanks to the Chinese-built Coca Codo Sinclair hydroelectric project, Ecuador no longer depends on imported energy. The project also created 15,000 jobs.

**15** In Guinea, the Chinese-built Kaleta Hydropower Station has greatly relieved power shortages in the capital, Conakry, and surrounding areas, benefiting 4 million people.

**16** Third, China has created a new channel for BRI countries to train water resource management specialists. We have carried out technical training for countries with relatively weak water resource management capabilities.

**17** At present, nearly 4,000 technicians and government officials from 112 countries have been provided with online and on-site training in English, French, Russian and Spanish. We have set up five overseas technology transfer and capacity building centers in Pakistan, Ethiopia, Indonesia, Serbia and Senegal.

**18** By focusing on building what we call "small but beautiful" projects, the ministry continues to promote the implementation of projects that change lives, such as those related to irrigation, flood control and water supply. It will also endeavor to strengthen the role of Chinese companies in offering design services overseas and promote international capacity cooperation. More efforts will also be made to provide more training in BRI countries to help them strengthen their water resource management capacity.

## Writing

Focus on one of the six aspects of President Xi's Thought on Ecological Civilization. Illustrate your understanding with specific examples.

## Unit Project

Organize a speech contest in your class. Give a presentation based on your writing.

Unit 7

# Agricultural Development and Management

Agriculture covers a broad scope of multidisciplinary and interdisciplinary fields that encompass the parts of natural, economic and social sciences. It is complex and sometimes also unpredictable. Moreover, there is increasing pressure for the development of agriculture due to the fast changes happening around the world from natural factors such as climate change, soil erosion, water pollution, etc. and social factors such as food scarcity, alterations of consumers' desires, etc. Solutions must be found to maintain the balance between the impacts of agriculture on the environment, climate, society, etc. and the impacts from the changes in climate, biodiversity and other environmental conditions on agriculture. Modern agriculture refers to the use of new technologies and agricultural practices to make food production more efficient, productive and sustainble. Germany and China, two major agricultural powers, find themselves at the forefront of addressing the current issues and charting a path towards a sustainable future. By recognizing their shared vulnerabilities and leveraging their respective strengths, Germany and China have the opportunity to collaborate and drive the transformation of agri-food systems, ensuring food security, environmental sustainability, and equitable outcomes for all.

*Practical English for Ecological Environment*

## Warm-up

You are going to watch a video about agriculture. Watch it twice and complete the following task. Compare your answers with your partner's.

① This means we can't _____, because doing so will _____ that make agriculture possible in the first place.

② Instead, the next agricultural revolution will have to increase the output of our existing farmland for the long term while _____, _____ and _____.

③ And in India, where up to 40 percent of post-harvest food is lost or wasted _____, farmers have already started to implement solar-powered cold storage capsules that help thousands of rural farmers _____ and become a viable part of the supply chain.

④ High-tech interventions stand to _____ to farming, and large producers will need to invest in _____.

⑤ If we optimize food production, both on land and sea, we can _____ within the environmental limits of the earth, but there's a very small _____, and it will take _____ of the agricultural lands we have today.

## Reading Comprehension

**TEXT A**

### Sino-German Agri-Cooperation for Mutual Benefit

Michaela Boehme  Jürgen Ritter

1  Agri-food systems worldwide are facing enormous challenges. And after steadily declining for a decade, world hunger is on the rise again, affecting nearly 10 percent of people globally.

2  At the same time, prices of farm inputs and food products are rising, exacerbating poverty and food insecurity among the world's most vulnerable populations. <u>As a result</u>, creating a

164

world free of hunger as set out in the 2030 Agenda for Sustainable Development of the United Nations appears to be an ever-more distant **prospect**.

3   There is no single reason behind the current challenges. Rather, we are facing multiple, interconnected crises—from climate change to loss of biodiversity and healthy soil, global **pandemics**, and wars and **conflicts** in many parts of the world.

4   In light of these challenges, the German Commission on the Future of Agriculture has developed pathways for a sustainable transformation of the country's agricultural sector. The proposed solutions range from creating more environmentally friendly livestock production systems to the promotion of organic agriculture, and the reduction of **fertilizers** and **pesticides**. Efforts at the level of the European Union, such as the EU's Green Deal or its Farm-to-Fork strategy, also aim to make food systems healthy and environmentally friendly, while ensuring productivity and fair returns to farmers.

5   In China, agricultural production has benefited from the fast **rollout** of modern agricultural machinery, digital tools, and innovations in biotechnology, thereby helping the country to raise agricultural productivity and grain output levels. With an annual output of more than 650 million tons, China has basically been self-sufficient in the production of staple grains for many years—achievement that has helped ease pressure on food supplies globally.

6   Yet neither Germany nor China is immune to the crises facing our food systems today. In fact, there are many shared challenges that risk undermining agricultural production in both countries in the medium term. For example, extreme weather events linked to climate change are becoming more frequent and severe. Regions in China and Germany were hit by major droughts in 2022. Protecting biodiversity and soil health is another issue of common concern. China's "zero-growth" action plan regulating the excessive use of fertilizers and pesticides is a step in the right direction, but more needs to be done.

7   In Germany, where fertilizer use is substantially lower than in China, farmers are now experimenting with new technologies to further improve **efficacy** and reduce environmental impacts.

8   Moreover, both countries are exploring the use of digital tools to tackle rural challenges such as a declining farming population and rural development. As two major agricultural powers, Germany and China have the joint responsibility to drive the transformation toward a more

sustainable, climate-friendly, and fair food system that can ensure food security for all while at the same time protecting our common global goods.

9    A constructive exchange and dialogue across national borders is **indispensable** to address the challenges we are facing, as parliamentary state secretary of the Federal Ministry of Food and Agriculture, Ophelia Nick, reiterated at the 8th Sino-German Agricultural Week, held from Nov. 21-25, 2022, in Beijing.

10    How do we best organize these exchanges? The Sino-German Agricultural Centre (DCZ), established in 2015 as a joint initiative of the agricultural ministries of both countries, plays a key role in facilitating necessary cooperation.

11    Designed to be a central contact and information platform, its multi-stakeholder approach has helped bring together actors from politics, science, and business from both countries to address common challenges and support the sustainable development of the agriculture and food industry.

12    Over the past few years, agricultural cooperation via the platform of the DCZ has addressed a wide range of issues. In 2019, a Sino-German cooperation project on agriculture and climate change was launched. Within the framework of the project, experts from the Chinese Academy of Agricultural Sciences, the Ministry of Agriculture and Rural Affairs, as well as German researchers were invited to study tours in each other's country.

13    Furthermore, a set of best practices and policy recommendations on how to reduce emissions from agriculture and **mitigate** the effects of climate change were developed. German and Chinese agri-food **stakeholders** have also **partnered** in the field of agricultural digitalization, with a smart agriculture working group set up in 2020 and the successful launch of a smart agriculture website, presenting insights from Germany's "Digital Experimental Fields".

14    Other cooperation activities have included in-depth professional exchanges on environmentally friendly management practices of animal manure, plant **breeding** and digital villages.

15    However, more can be done. Balancing food production goals with the need to protect our global public goods, including our ecosystems and climate, is one of the biggest challenges facing both Germany and China in the near future. Addressing this challenge will require a wide range of responses, from innovation in digital agriculture to the repositioning of subsidies, the promotion of regenerative farming practices, and the creation of a fair and stable trade environment.

16  By drawing on their respective experiences, Germany and China can and should develop solutions for jointly overcoming the multiple crises and contribute to building a fair, healthy and sustainable food system for all.

## New Words

**prospect** [ˈprɒspekt] *n.*
reasonable hope of sth happening; sth which is probable soon 前景；前途
**pandemic** [pænˈdemɪk] *n.*
an epidemic that is geographically widespread; occurring throughout a region or even throughout the world 流行病
**conflict** [ˈkɒnflɪkt] *n.*
an open clash between two opposing groups (or individuals); struggle; fight 战斗；斗争；冲突
**fertilizer** [ˈfɜːtəlaɪzə(r)] *n.*
natural or artificial substance added to soil to make it more fertile 肥料；化肥
**pesticide** [ˈpestɪsaɪd] *n.*
substance for killing insects 杀虫剂
**rollout** [ˈrəlˌaʊt] *n.*
the act of making something, especially a product or service, available for the first time （新产品或服务的）推出；首次展示
**efficacy** [ˈefɪkəsi] *n.*
capacity or power to produce a desired effect 效能；功效
**indispensable** [ˌɪndɪˈspensəbl] *adj.*
absolutely necessary; vitally necessary 不可缺少的
**mitigate** [ˈmɪtɪgeɪt] *v.*
lessen or to try to lessen the seriousness or extent of 减轻；（使）缓和
**stakeholder** [ˈsteɪkhəʊldə(r)] *n.*
someone entrusted to hold the stakes for two or more persons betting against one another; must deliver the stakes to the winner 利益相关者
**partner** [ˈpɑːtnə(r)] *vt.*
act as or be the partner of sb 做……的搭档
**breed** [briːd] *v.*
give birth to young; reproduce; keep animals, plants, etc. 生育；繁殖；饲养

## Phrases and Expressions

**as a result:** consequently; therefore 结果是；因此

**set out:** to explain, describe, or arrange something in a clear and detailed way, especially in writing; to start doing or working on something in order to achieve a goal 陈述；计划

**in light of:** according to 按照；根据；鉴于

**range from...to**: to vary between two particular amounts, sizes, etc., including others between them（在一定的范围内）变化，变动

**be immune to:** not influenced or affected by something 不受影响

**set up:** establish 建立，设立

**draw on**：to use (information, experience, knowledge, etc.) to make something 利用，运用（信息、经验、知识等）

### Exercises

**I Understanding the Text**

1. Recognize the sequence in which the passage presents its most important information on Sino-German agri-cooperation. Fill in the brackets with the correct choice form the following list.

① Both German and China have made great achievements in agriculture.

② Cooperation activities have been carried out between German and China in various aspects.

③ The writer draws a conclusion and reiterates his main idea.

④ Both German and China are facing crises in food systems and have many shared challenges.

⑤ Agri-food systems worldwide are facing enormous challenges.

| Parts | Paragraphs | Main ideas |
| --- | --- | --- |
| Part One | Paras.1-3 | |
| Part Two | Paras.4-5 | |
| Part Three | Paras.6-8 | |
| Part Four | Paras.9-15 | |
| Part Five | Para.16 | |

2. Focus on the sentences that express causes and effects. Pay attention to the sentence patterns and fill in key information.

① And after steadily declining for a decade, world hunger is on the rise again, _____.

② At the same time, prices of farm inputs and food products are rising, _____ among the world's most vulnerable populations.

③ _____ as set out in the 2030 Agenda for Sustainable Development of the United Nations appears to be an ever-more distant prospect.

④ _____ behind the current challenges.

⑤ _____, the German Commission on the Future of Agriculture _____ of the country's agricultural sector.

⑥ In China, agricultural production has benefited from the fast rollout of modern agricultural machinery, digital tools, and innovations in biotechnology, _____.

**II Focusing on Language in Context**

1. Key Words & Expressions

A. Fill in the blanks with the words given below. Change the form where necessary. Each word can be used only once.

**mitigate breed partner stakeholders indispensable efficacy rollout prospect pesticide conflict**

① They have to _____ with the government, even if they are dismissive of its bureaucracy.

② Air is _____ to life.

③ Wild animals don't _____ well in captivity.

④ He's got involved in a political _____ he can't extricate himself from.

⑤ A rich harvest is in _____.

⑥ Reporting accurate figures to _____ is crucial for gaining public trust.

⑦ We are testing the _____ of a new drug.

⑧ The government is trying to _____ the effects of inflation.

⑨ Since its _____ in fall of 1999, the online service has gained millions of members.

⑩ The insects have become resistant to the _____.

B. Fill in the blanks with the phrases given below. Change the form where necessary.

**set out; as a result; in light of; range from…to; be immune to; set up; draw on**

① When they _____, they were well prepared.

② The company is capable of supplying monodisperse particles of any size in the _____ 5 nanometers _____ 1000 microns.

③ It seems no one _____ vanity, just as no one can survive without food.

④ _____, we have to water the vegetable garden.

⑤ They _____ their previous experience.

⑥ A fund will be _____ for the dead men's families.

⑦ Reconsider the questions _____ what you know now and try to write down the group decision.

2. Usage

A. Complete the following sentences in the text by using elliptical sentences.

169

① _____, world hunger is on the rise again, affecting nearly 10 percent of people globally.（并且，在连续十年稳步下降之后）

② Efforts at the level of the European Union, such as the EU's Green Deal or its Farm-to-Fork strategy, also aim to make food systems healthy and environmentally friendly, _____ _____.（同时确保生产力和农民的公平收益）

③ As two major agricultural powers, Germany and China have the joint responsibility to drive the transformation toward a more sustainable, climate-friendly, and fair food system that can ensure food security for all _____.（同时保护我们共同的全球利益）

B. Complete the following sentences in the text by using "from...to".

① Rather, we are facing multiple, interconnected crises—_____ _____ in many parts of the world.（从气候变化到生物多样性和健康土壤的消失，从全球大流行病到战争和冲突）

② The proposed solutions _____.（从建立更环保的牲畜生产系统到促进有机农业，以及减少化肥和农药的使用）

**III Translation**

Translate the following passage into English.

近年来，我国也加强了在应对气候变化和环境保护方面的工作力度。特别是在农业领域，2016年10月中国提出了《全国农业现代化规划》。除常规增加农业总产量目标外，计划在畜牧业和农作物生产中使用环境友好型方法。德国农业在应对气候变化方面出台的政策和技术对于减少二氧化碳排放、减缓和适应气候变化发挥了较好的作用，值得我国学习借鉴，主要是进一步加强科技研究和进一步加强政策创设。

**IV Pair Work**

Talk about the contributions China has made to food production and world food security.

**TEXT B**

微课

## Green Unity

Zhang Xiuqing

1  China's grain production has been <u>on the rise</u>. The annual grain output remained stable at over 650 million metric tons over the past eight years and reached 687 million metric tons in 2022,

a record high. However, as the economy grows and there is greater urbanization, increasing agricultural production to meet the rising demands for grain and important agricultural sideline products remains a challenging task.

2  In addition, the agricultural sector also contributes to global warming. Farming, using fertilizers in farmland and breeding **livestock** generate greenhouse gasses. <u>According to studies, greenhouse gas emissions from agriculture, which mainly come from plantations, animal **husbandry** and waste treatment, account for 7-15 percent of China's total greenhouse gas emissions.</u>

3  With a large population, relatively small area of **arable** land, and scarce water resources, China has been relying on massive investment of resources and heavy use of fertilizers and water to increase crop yield and speed of harvest, which has resulted in serious environmental problems such as soil **compaction**, soil **degradation** and pollution from pesticides and fertilizers. Such production methods are no longer sustainable.

4  <u>To meet China's goals of peaking its carbon dioxide emissions before 2030 and achieving carbon **neutrality** before 2060, on the one hand, a green transformation of agriculture production should be promoted.</u> Instead of pursuing high **yields**, we should <u>strive to</u> achieve stable production and supply while reducing greenhouse gas emissions.

5  On the other hand, we need to <u>tap into</u> the "green" nature of agriculture, an important component of the ecosystem. Rice fields, vegetable farms and **orchards** are green spaces and wetlands, which can provide carbon **sequestration** and carbon sink functions.

6  By combining the economic value of agricultural products with the ecological value of agriculture and rural areas, China can achieve sustainable and high-quality development of its economy and society.

7  To achieve stable and increased agricultural production under the carbon peaking and carbon neutrality goals, China must adopt measures to encourage carbon reduction in the production process, promote green and efficient **circulation**, and reduce food wastage at the consumption end.

8  <u>At the same time, the government needs to **motivate** agricultural production **entities** to go for green development and **integrate** national food security with green and low-carbon transformation.</u>

9  In fact, a carbon reduction pressure transmission mechanism is forming in the agricultural industry chain, where agricultural companies usually play a leading role, connecting smallholders

and large markets. For companies, the pressure to reduce carbon emissions comes not only from macro policies but also from upstream and downstream enterprises in the industrial chain.

**10** Currently, more and more companies are starting to **disclose** their environmental information to the public.

**11** According to data from the Carbon Disclosure Project, a global environmental information research center, the number of Chinese companies that **submit** relevant environment information to it exceeded 1,300 in 2020. And 11 companies received A or A—ratings from the center, which is the best level in history. Also, in China's A-share stock **bourse**, the number of companies that publish environmental social and governance-related reports increased from 872 in 2018 to 1,130 in 2021.

**12** Downstream companies that disclose their environmental information will inevitably demand upstream suppliers to meet green standards in areas such as their agricultural products' planting environment, transportation process and packaging. This pressure will transmit along agribusinesses in the industry chain.

**13** For example, Yili Group is one of the top five dairy producers in the world, and 90 percent of Yili Group's carbon emissions come from its activities throughout the supply chain, such as raw material **procurement**, packaging, and logistics. Yili is currently working on sorting out emission sources and exploring the best method for carbon emissions calculation.

**14** The downstream enterprises' demands for green **attributes** for their products will be transmitted to small-scale agricultural production entities at the front end of the industry chain.

**15** The government needs to pay attention to this transmission mechanism and give good guidance and effective measures to **cultivate** its development.

**16** First, to build a whole-process monitoring system to **safeguard** food supply while reducing carbon generation in agriculture,. Information on the process of agricultural production, circulation, consumption, environmental protection, and carbon reduction needs to be gathered and monitored using big data technology.

17  Second, green agricultural taxation and insurance policies need to be improved, so as to support and encourage green agricultural investment. For example, policies such as giving companies tax breaks for the production of green and low-carbon agricultural goods or tax deductions for the purchase of low-carbon equipment can be introduced. Also, insurance companies can be encouraged to play a greater role in responding to the risk management needs of agricultural companies in their green transformation.

18  <u>Third, there should be an agricultural carbon sequestration compensation mechanism to provide economic **incentives** for agricultural entities.</u> By designing scientific measurement and calculation standards for agricultural carbon sequestration, as well as formulating compensation principles, special subsidies can be issued and the carbon market can **incentivize** the transformation to green agriculture.

19  Finally, China should strengthen international cooperation in agricultural science and technology and carbon reduction while sharing China's good practices and technologies. At the same time, we need to attract global talents in carbon reduction, create more international cooperation, innovation and sharing platforms, and improve the mechanisms for exchanges, cooperation, and sharing of international green agricultural technologies.

### New Words

**livestock** [ˈlaɪvstɒk] *n.*
farm animals;domestic animals kept for use or their value as a source of food and other products 家畜，牲畜

**husbandry** [ˈhʌzbəndri] *n.*
the practice of cultivating the land or raising stock 农事，饲养业

**arable** [ˈærəbl] *adj.*
(of farmland) capable of being farmed productively 可耕的；适合种植的

**compaction** [kəmˈpækʃən] *n.*
an increase in the density of something 压缩

**degradation** [ˌdeɡrəˈdeɪʃn] *n.*
changing to a lower state (a less respected state); a low or downcast state 降格；堕落；退化

**neutrality** [njuːˈtræləti] *n.*
nonparticipation in a dispute or war; tolerance attributable to a lack of involvement 中立

**yield** [jiːld] *n.*
that which is yielded or produced 产量，收益

**orchard** [ˈɔːtʃəd] *n.*

(usually enclosed) piece of land in which fruit trees are grown 果园

**sequestration** [ˌsiːkwəˈstreɪʃn] *n.*

the action of forming a chelate or other stable compound with an ion or atom or molecule so that it is no longer available for reactions 隐退；隔离

**circulation** [ˌsɜːkjəˈleɪʃn] *n.*

the flow of gas or liquid around a closed system 流通；循环

**motivate** [ˈməʊtɪveɪt] *v.*

inspire; influence; incite; stimulate 激发；诱发

**entity** [ˈentəti] *n.*

a single separate and independent existence 实体；实际存在物

**integrate** [ˈɪntɪgreɪt] *v.*

make into a whole or make part of a whole 整合；结合

**disclose** [dɪsˈkləʊz] *v.*

make known to the public information that was previously known only to a few people or that was meant to be kept a secret 揭露；公开

**submit** [səbˈmɪt] *v.*

put forward for opinion, discussion, decision, etc. 提出

**bourse** [bʊəs] *n.*

the stock exchange 证券交易所

**procurement** [prəˈkjʊəmənt] *n.*

the act of getting possession of something 采购；获得

**attributes** [ˈætrɪbjuːt] *n.*

a quality forming part of the nature of a person or thing 属性；特性

**cultivate** [ˈkʌltɪveɪt] *v.*

prepare and use (land, soil, etc.) for growing crops 耕作；种植

**safeguard** [ˈseɪfɡɑːd] *v.*

protect or guard 保护；保卫

**incentive** [ɪnˈsentɪv] *n.*

a positive motivational influence 激励某人做某事的事物

**incentivize** [ɪnˈsentɪvaɪz] *v.*

to make someone want to do something 刺激；激励

## Phrases and Expressions

**on the rise:** increasing in amount, number, level, etc.（数量、水平等）增长，上升

**strive to:** to try hard to do something 争取；追求

**tap into:** to use energy, information, money etc. that comes from a large supply 深入了解；（为了利益）充分利用或挖掘（资源）

**sort out:** arrange or order by classes or categories 整理；分类

### Exercises

**I Comprehension Check**

① What makes agricultural production remain a challenging task?

② Why does the agricultural sector contribute to global warming?

③ What are China's goals?

④ What are the sustainable production methods?

⑤ What can the government do to cultivate the development?

**II Translation**

Translate into Chinese the underlined sentences in the article.

① According to studies, greenhouse gas emissions from agriculture, which mainly come from plantations, animal husbandry and waste treatment, account for 7-15 percent of China's total greenhouse gas emissions.

② To meet China's goals of peaking its carbon dioxide emissions before 2030 and achieving carbon neutrality before 2060, on the one hand, a green transformation of agriculture production should be promoted.

③ At the same time, the government needs to motivate agricultural production entities to go for green development and integrate national food security with green and low-carbon transformation.

④ Currently, more and more companies are starting to disclose their environmental information to the public.

⑤ Third, there should be an agricultural carbon sequestration compensation mechanism to provide economic incentives for agricultural entities.

**III Group Work**

Have you ever been to an agritainment resort? Why is agri-tourism booming and what are the benefits?

**TEXT C**

## Agriculture's Connected Future: How Technology Can Yield New Growth

Lutz Goedde, Joshua Katz, Alexandre Ménard, and Julien Revellat

(One of the oldest industries must embrace a digital, connectivity-fueled transformation in order to overcome increasing demand and several disruptive forces.)

1  The agriculture industry has radically transformed over the past 50 years. Advances in machinery have expanded the scale, speed, and productivity of farm equipment, leading to more efficient cultivation of more land. Seed, irrigation, and fertilizers also have vastly improved, helping farmers increase yields. Now, agriculture is in the early days of yet another revolution, at the heart of which lie data and connectivity. Artificial intelligence, analytics, connected sensors, and other emerging technologies could further increase yields, improve the efficiency of water and other inputs, and build sustainability and **resilience** across crop cultivation and animal husbandry.

2  Without a solid connectivity infrastructure, however, none of this is possible. If connectivity is implemented successfully in agriculture, the industry could tack on $500 billion in additional value to the global gross domestic product by 2030, according to our research. This would amount to a 7 to 9 percent improvement from its expected total and would **alleviate** ease much of the present pressure on farmers. It is one of just seven sectors that, **fueled** by advanced connectivity, will contribute $2 trillion to $3 trillion in additional value to global GDP over the next decade, according to research by the McKinsey Center for Advanced Connectivity and the McKinsey Global Institute (MGI) (see sidebar "The future of connectivity").

3  Demand for food is growing at the same time the supply side faces **constraints** in land and farming inputs. The world's population is on track to reach 9.7 billion by 2050, requiring a corresponding 70 percent increase in calories available for consumption, even as the cost of

the inputs needed to generate those calories is rising. By 2030, the water supply will fall 40 percent short of meeting global water needs, and rising energy, labor, and nutrient costs are already **pressuring** profit margins. About one-quarter of arable land is degraded and needs significant **restoration** before it can again sustain crops <u>at scale</u>. And then there are increasing environmental pressures, such as climate change and the economic impact of **catastrophic** weather  events, and social pressures, including the push for more ethical and sustainable farm practices, such as higher standards for farm-animal welfare and reduced use of chemicals and water.

4   To address these forces **poised** to further **roil** the industry, agriculture must embrace a digital transformation enabled by connectivity. Yet agriculture remains less digitized compared with many other industries globally. Past advances were mostly mechanical, in the form of more powerful and efficient machinery, and genetic, in the form of more productive seeds and fertilizers. Now much more sophisticated, digital tools are needed to deliver the next productivity leap. Some already exist to help farmers more efficiently and sustainably use resources, while more advanced ones are <u>in development</u>. These new technologies can upgrade decision making, allowing better risk and variability management to optimize yields and improve economics. Deployed in animal husbandry, they can enhance the well-being of livestock, addressing the growing concerns over animal welfare.

5   But the industry **confronts** two significant obstacles. Some regions lack the necessary connectivity infrastructure, making development of it **paramount**. In regions that already have a connectivity infrastructure, farms have been slow to deploy digital tools because their impact has not been sufficiently proven.

6   <u>The COVID-19 crisis has further intensified other challenges agriculture faces in five areas: efficiency, resilience, digitization, **agility**, and sustainability.</u> Lower sales volumes have pressured margins, exacerbating the need for farmers to contain costs further. Gridlocked global supply chains have highlighted the importance of having more local providers, which could increase the resilience of smaller farms. In this  global pandemic, heavy reliance on manual labor has further affected farms whose workforces

face mobility restrictions. Additionally, significant environmental benefits from decreased travel and consumption during the crisis are likely to drive a desire for more local, sustainable sourcing, requiring producers to adjust long-standing practices. In short, the crisis has **accentuated** the necessity of more widespread digitization and automation, while suddenly shifting demand and sales channels have **underscored** the value of **agile** adaptation.

### Current connectivity in agriculture

7  In recent years, many farmers have begun to consult data about essential variables like soil, crops, livestock, and weather. Yet few if any have had access to advanced digital tools that would help to turn these data into valuable, actionable insights. In less-developed regions, almost all farmwork is manual, involving little or no advanced connectivity or equipment.

8  Even in the United States, a pioneer country in connectivity, only about one-quarter of farms currently use any connected equipment or devices to access data, and that technology isn't exactly state-of-the-art, running on 2G or 3G networks that telcos plan to **dismantle** or on very low-band IoT networks that are complicated and expensive to set up. In either case, those networks can support only a limited number of devices and lack the performance for real-time data transfer, which is essential to unlock the value of more advanced and complex use cases.

9  Nonetheless, current IoT technologies running on 3G and 4G cellular networks are in many cases sufficient to enable simpler use cases, such as advanced monitoring of crops and livestock. In the past, however, the cost of hardware was high, so the business case for implementing IoT in farming did not <u>hold up</u>. Today, device and hardware costs are dropping rapidly, and several providers now offer solutions at a price we believe will deliver a return in the first year of investment.

10  These simpler tools are not enough, though, to unlock all the potential value that connectivity holds for agriculture. To attain that, the industry must make full use of digital applications and analytics, which will require low **latency**, high bandwidth, high resiliency, and support for a density of devices offered by advanced and frontier connectivity technologies like LPWAN, 5G, and LEO satellites.

11  The challenge the industry is facing is thus twofold: infrastructure must be developed to enable the use of connectivity in farming, and where connectivity already exists, strong business cases must be made in order for solutions to be adopted. The good news is that connectivity coverage is increasing almost everywhere. By 2030, we expect advanced connectivity infrastructure of some type to cover roughly 80 percent of the world's rural areas; the notable exception is Africa, where only a quarter of its area will be covered. The key, then, is to develop more—and more effective—digital tools for the industry and to foster widespread adoption of them.

**12** As connectivity increasingly takes hold, these tools will enable new capabilities in agriculture:

• Massive Internet of Things. Low-power networks and cheaper sensors will set the stage for the IoT to scale up, enabling such use cases as precision irrigation of field crops, monitoring of large herds of livestock, and tracking of the use and performance of remote buildings and large fleets of machinery.

• Mission-critical services. Ultralow latency and improved stability of connections will foster confidence to run applications that demand absolute reliability and responsiveness, such as operating autonomous machinery and drones.

• Near-global coverage. If LEO satellites attain their potential, they will enable even the most remote rural areas of the world to use extensive digitization, which will enhance global farming productivity.

**Connectivity's potential for value creation**

**13** By the end of the decade, enhanced connectivity in agriculture could add more than $500 billion to global gross domestic product, a critical productivity improvement of 7 to 9 percent for the industry.5 Much of that value, however, will require investments in connectivity that today are largely absent from agriculture. Other industries already use technologies like LPWAN, cloud computing, and cheaper, better sensors requiring minimal hardware, which can significantly reduce the necessary investment. We have analyzed five use cases—crop monitoring, livestock monitoring, building and equipment management, drone farming, and autonomous farming machinery—where enhanced connectivity is already in the early stages of being used and is most likely to deliver the higher yields, lower costs, and greater resilience and sustainability that the industry needs to thrive in the 21st century .

**14** It's important to note that use cases do not apply equally across regions. For example, in North America, where yields are already fairly optimized, monitoring solutions do not have the same potential for value creation as in Asia or Africa, where there is much more room to improve productivity. Drones and autonomous machinery will deliver more impact to advanced markets, as technology will likely be more readily available there .

**15** Agriculture, one of the world's oldest industries, finds itself at a technological crossroads. To handle increasing demand and several **disruptive** trends successfully, the industry will need to overcome the challenges to deploying advanced connectivity. This will require significant investment in infrastructure and a **realignment** of traditional roles. It is a huge but critical **undertaking**, with more than $500 billion in value at stake. The success and sustainability of one of the planet's oldest industries may well depend on this technology transformation, and those that embrace it at the outset may be best positioned to thrive in agriculture's connectivity-driven future.

### New Words

**resilience** [rɪˈzɪliəns] *n.*

the physical property of a material that can return to its original shape or position after deformation that does not exceed its elastic limit 适应力

**alleviate** [əˈliːvieɪt] *v.*

make (sth) less severe; ease 减轻；缓解；缓和

**fuel** [ˈfjuːəl] *v.*

to give support or strength to (something) 加强；增强

**constraint** [kənˈstreɪnt] *n.*

the threat or use of force as a strong influence on one's actions 限制，约束

**pressure** [ˈpreʃə(r)] *v.*

to cause to do through pressure or necessity, by physical, moral or intellectual means 施压

**restoration** [ˌrestəˈreɪʃn] *n.*

the act of restoring something or someone to a satisfactory state 恢复

**catastrophic** [ˌkætəˈstrɒfɪk] *adj.*

extremely harmful; bringing physical or financial ruin 灾难的；灾难性的

**poise** [pɔɪz] *v.*

prepare (oneself) for something unpleasant or difficult 做准备

**roil** [rɔɪl] *v.*

be agitated; make turbid by stirring up the sediments of 搅浑；激怒；动荡

**confront** [kənˈfrʌnt] *v.*

bring face to face; be or come face to face with 对抗；遭遇；面临

**paramount** [ˈpærəmaʊnt] *adj.*

having superior power and influence 极为重要的；至高无上的

**agility** [əˈdʒɪləti] *n.*

the gracefulness of a person or animal that is quick and nimble 敏捷；灵活；轻快

**accentuate** [əkˈsentʃueɪt] *v.*

to stress, single out as important 强调；突出

**underscore** [ˌʌndəˈskɔː(r)] *v.*

give extra weight to (a communication) 强调

**agile** [ˈædʒaɪl] *adj.*

moving quickly and lightly（动作）敏捷的；灵活的

**dismantle** [dɪsˈmæntl] *v.*

bring to an end (a system, arrangement, etc.), especially by gradual stages 废除，取消

**latency** [ˈleɪtənsi] *n.*

the time that elapses between a stimulus and the response to it 延迟

**disruptive** [dɪsˈrʌptɪv] *adj.*

characterized by unrest or disorder or insubordination 破坏性的

**realignment** [ˌriːəˈlaɪnmənt] *n.*

the act of changing your opinions, policies, etc. so that they are the same as those of another person, group, etc. 改变观点，改变策略（以与别人相同）

**undertaking** [ˌʌndəˈteɪkɪŋ] *n.*

any piece of work that is undertaken or attempted 事业

**tack on:** to add something that you had not planned to add, often without much preparation or thought 添加；附加

**amount to:** add up to 总计

**at scale:** on a large scale 大规模

**in development**：the state of being created or made more advanced 在发展中；在开发中

**hold up**：to support someone or something so that they do not fall down 支撑

**take hold:** to become effective, established, or popular 变得有效；得以建立

**scale up:** to **make something larger in size, amount, extent etc. than it used to be** 扩大，增加

**at a crossroads:** at an important point in somebody's life or development （人生或发展）处于关键时刻；在紧要关头

**at the outset:** at first; initially 在开头；开始时

## Exercises

**Translation**

Translate into Chinese the underlined sentences in the article

① One of the oldest industries must embrace a digital, connectivity-fueled transformation in order to overcome increasing demand and several disruptive forces.

② Advances in machinery have expanded the scale, speed, and productivity of farm equipment, leading to more efficient cultivation of more land.

③ Without a solid connectivity infrastructure, however, none of this is possible.

④ The COVID-19 crisis has further intensified other challenges agriculture faces in five areas: efficiency, resilience, digitization, agility, and sustainability.

⑤ The success and sustainability of one of the planet's oldest industries may well depend on this technology transformation, and those that embrace it at the outset may be best positioned to thrive in agriculture's connectivity-driven future.

## Supplementary Reading

**TEXT A**

### Should You Buy Organic? "Dirty Dozen" Pesticide List Helps You Decide

1  If strawberries, spinach or kale is on your shopping list, you might want to opt for the organic versions the next time you are at the grocery store. The three items were found to have the highest levels of pesticide residue, according to an annual analysis from the Environmental Working Group (EWG).

2  The annual EWG list ranks popular fruits and vegetables based on levels of contamination. The resulting lists—the "dirty dozen" and "clean 15"—outline the produce found to have the highest and lowest levels of pesticide contamination in the United States.

3  Strawberries topped this year's dirty dozen list for the fourth year in a row. Kale—a nutrient-rich cabbage—was a surprising newcomer at number three on the list. Ninety-two percent of the kale samples had two or more pesticide residues. Nectarines, apples, and peaches were also on the "dirty" list.

4  At the top of the "clean 15" list are avocado, sweet corn and pineapple. Popular items such as onions, mushrooms and cabbage were also among the cleanest produce items. "More than 70 percent of 'clean 15' fruit and vegetable samples had no pesticide residues", said the report.

5  Pesticides are commonly used to control weeds, insects or unwanted vegetation in agriculture. People are exposed to pesticides through the food they eat because of their broad use.

6  The World Health organization says pesticides can be hazardous and have toxic effects. However, scientists still do not have a clear understanding of the health effects of pesticide residues, writes the National Institute of Environmental Health Science.

7  Rankings are based on an analysis of more than 40,000 samples taken by the United States Department of Agriculture (USDA) and the U.S. Food and Drug Administration (FDA).

8  EWA says the report is intended to help shoppers decide when to buy organic versus conventional.

9  The report cites a French study published in *JAMA Internal Medicine* last year that found that among almost 69,000 participants, those who ate organic more often had 25 percent fewer cancers than those who didn't eat organic food.

10  Because the items on the clean list have low pesticide residue, shoppers can buy standard items, while organic might be a better option for produce on the dirty dozen list. Recognizing this isn't always an option for people, the report acknowledges that eating any fresh fruit or vegetable—organic or not—is beneficial to one's health.

## TEXT B

### Biomass Technology Shows Huge Growth

Zheng Xin

1  Organic waste, including wood, crop byprodu[...] an important role in facilitating global carbon neutrality[...] cars.

2  In Heilongjiang Province, State Power Investme[...] to compress corn straw, residues and agricultural and [...] provide clean heating to local residents.

3  The technology will be put into use by 2024 and replace coal to provide clean heating for more than 10 million square meters in Jiamusi, Heilongjiang, the company said.

4  As the country's first domestic biomass green energy particle technology, it will make better use of the availability of large quantities of corn straw and other residues, breaking the bottleneck of inconvenient transportation and storage for biomass energy utilization, said Ma Mingjun, general manager of Shanghai Power Equipment Research Institute Co Ltd.

5  Biomass, such as agricultural and forestry products, organic household waste as well as livestock and industrial refuse, are some of the biological materials used as fuels in producing electricity and heat. It can be burned directly for heat or converted to renewable fuels through thermal, chemical and biochemical processes.

6  Under the energy efficiency and carbon intensity targets set by the Chinese government, the country's development of biomass energy is likely to be fast-tracked thanks to preferential policy support, experts said.

7  China's development of the biomass energy industry is set to embrace major opportunities under China's strong green commitment, according to Zhang Dayong, secretary-general of China's Biomass Energy Industry Promotion Association.

8  The industry has great potential for further growth as China strives to achieve carbon peak by 2030 and carbon neutrality by 2060, he said.

9  China produces over 900 million metric tons of agricultural and forestry biomass every year, which can generate power equal to nearly 400 million tons of coal. The number is even larger including other organic waste from urban and rural areas, according to the association.

10  However, at present, only 90 million tons of agricultural and forestry biomass is used for power generation annually. The high costs of collecting raw materials and relatively low power generation rates compared with coal and other mainstream energy sources have been hindering the industry's development, officials said.

11  State Power Investment Corp's attempt to compress corn straw, understory residues and agricultural and related processing wastes into fuel, however, is expected to represent a breakthrough for biomass heating in the country, said Luo Zuoxian, head of intelligence and research at the Sinopec Economics and Development Research Institute.

12  The country's 14th Five-Year Plan (2021—2025) has sent positive signals encouraging biomass energy, he said.

13  The country's installed capacity for biomass energy rose to 37.98 million kilowatts by the end of last year, while the annual power generation capacity for biomass also rose to 163.7 billion kilowatt-hours during the same period, according to the National Energy Administration.

14  Last year, China's installed capacity of biomass power generation connected to the grid increased by 8.08 million kW, a record high that also ranks first in such field in the world, it said.

15  The administration has called for support from local governments for biomass energy projects, with heating being a priority.

16  China's strong green commitment will provide more opportunities for the growth of biomass energy development as biomass is a net zero-carbon fuel compared with other renewable energy sources, Zhang said.

17  While burning biomass releases carbon dioxide, the plants that make up biomass capture almost the same amount of carbon dioxide while growing, experts said.

18  According to a report released by the association, the government is expected to provide more support to boost the industry's development in the next five years. An estimated 1.2 trillion yuan ($172.32 billion) is to be invested in the industry from 2021 to 2025. That is expected to help the industry handle about 350 million tons of organic waste and create job opportunities for around 1 million people, the report said.

19  By 2030, the proportion of biomass energy in renewable energy is forecast to increase to about 8 percent, it said.

20  The government is also working on the combination of biomass heating with carbon capture and storage. It will extract energy from biomass, capture and store the carbon and turn biomass into energy to achieve negative emissions, the association said.

## TEXT C

### Township Turns Waste Straw into Profit

Cang Wei

1  Excess straw used to pose a big headache for the town of Guanshan in Suining County, Jiangsu Province, but the locals are now selling the agricultural waste for money.

2  In the past, officials of the township's Sanlie Village patrolled farmland every year to prevent farmers from burning straw, which seriously polluted

the air. They also stopped farmers from throwing straw into brooks and ditches, which might cause water pollution.

3   Some farmers chose to let the straw decompose in the farmland, but germs in the straw can pollute the soil and harm crops. Even if the straw is not contaminated, a hectare of farmland can only decompose up to 1.5 metric tons of straw.

4   "Our village has 430 hectares of farmland and produces 1,700 metric tons of wheat straw a year," said Ge Xingfei, deputy head of Sanlie Village. "We now sell the straw and hire villagers to guard it."

5   Straw has not only been used to feed cattle, generate electricity and produce methane, but has also been turned into nutritious organic material for growing mushrooms.

6   Ge said a mushroom company that can consume 60,000 metric tons of straw and 60,000 tons of chicken manure has been invited to invest in the village.

7   "The Zhongyou Xinghe Mushroom company buys the straw at 500 yuan ($77) a ton, and the local government gives a subsidy of 80 yuan a ton," Ge said. "Apart from delivery and transportation costs, we can earn 280 yuan by selling 1 ton of straw."

8   Zhu Nixian, 62, is one of the many villagers in the township hired to guard the straw. She is paid 500 yuan a month, which guarantees her basic livelihood in the rural area.

9   "I'm happy I can still work," said Zhu, who has no pension or children. "I feel that I'm still needed, and the money can improve my life."

10   The mushroom residues are turned into nutrient-rich soil for growing flowers, said Cheng Zhen, director of Xuzhou Xingmida Trading.

11   "We noticed that mushroom residue was scattered in the corners when we visited the Zhongyou Xinghe company," he said. "It occurred to us that the residue could be made into nutrient-rich soil. After repeated experiments, we succeeded in producing the soil and started to sell it on the internet.

12   "Now our company's annual sales have surpassed 10 million yuan. We've hired more than 30 local women from nearby villages to work in the industry. The women, all elderly, can earn a monthly salary of more than 3,000 yuan."

**13** Guanshan is now home to nearly 20 companies that, between them, sell about 15,000 tons of nutrient-rich soil made from 30,000 tons of mushroom residue.

**14** The remaining 90,000 tons of residue are used as raw material for producing organic fertilizer, the Suining County information office said.

**15** An important agricultural county in Xuzhou, Suining cultivates crops on more than 147,000 hectares of land, producing more than 1 million tons of wheat and paddy straw every year.

**16** The county now has nearly 800 straw agents and 48 social organizations specialized in straw storage and utilization, 11 of which can make use of more than 10,000 tons of straw.

**17** It has established 37 methane production centers, which can bring in revenue of 130 million yuan a year. The methane they produce can meet the demand of 50,000 households.

**18** "Ecological circular agriculture has become a new engine for developing rural areas," said Su Wei, Party secretary of Suining County. "Suining has changed from a polluted smoggy area to a scenic county with numerous green trees and blue sky.

**19** "The county will continue to balance the development of the economy and ecology for both the companies and local residents, and will strive to develop ecological agriculture for a better future."

## Writing

What do you think of organic farming and food? Try to illustrate your ideas with specific examples. What measures can be taken to boost the green agriculture?

## Unit Project

Please make an investigation of the main crops of your hometown. Discuss the reasons why your hometown is suitable for them to grow. (Hints: soil condition, weather condition, etc.)

# Unit 8

# Dual Carbon Targets

Climate change is a serious issue of concern around the world. In October 2018, the Intergovernmental Panel on Climate Change (IPCC) proposed the carbon neutrality target to keep the increase of the atmosphere temperature below 1.5 ℃. All countries and parties should take actions. China actively participates in this global climate change action and fulfills its international obligations. Since September 2020, China has been committed to the "double carbon targets", with a two-step plan: first, China will strive to peak $CO_2$ emissions before 2030; second, China will achieve carbon neutrality before 2060. This framework explores the fundamental concepts associated with the dual carbon goals and China's current situation. Besides, it reveals the strategies of achieving dual carbon goals in order to obtain a general understanding of different approaches, such as carbon capture projects and forest sinks.

## Practical English for Ecological Environment

## Warm-up

**Task 1**  Watch the video and take notes according to the clues given.

（1）What is carbon neutrality?
（2）How is carbon neutrality reached?
（3）How can we compensate for our emissions of greenhouse gases?
（4）Is climate change a global problem or a regional problem?
（5）What is the main contributor to climate change?

**Task 2**  Watch the video again and fill in the blanks with the missing information. You can refer to the notes you have taken.

（1）The relationship between emissions of greenhouse gases and climate change is a _____ _____.

（2）Carbon neutrality reduces the _____ impact on the environment to achieve a net result of _____.

（3）First, with measures that reduce _____ as much as possible, improve _____, innovation in _____ technology, increase the consumption of electricity from _____, and eliminate the use of _____ that cause climate change.

（4）Using _____ where the sale of credits and rights between countries, organizations and individuals takes place, in order to fulfill the agreement reached at the climate summit in Paris.

（5）Climate change is a _____, we must think of global emissions to act _____ and _____.

（6）The way we _____ is the main contributor to climate change and carbon neutrality helps to change our habits.

**Task 3**  Work in small groups. Discuss with your group members the following questions.

We use energy all the time. Do you know where the energy comes from? Are you able to give some examples of the sources and uses of different types of energy? Do you know the differences between renewable sources and nonrenewable sources?

## Reading Comprehension

TEXT A

### Ushering in the Green Transformation

Chheang Vannarith

(ASEAN-China cooperation to promote sustainable development can help set an example for the world.)

1  The urgent need for green development has gained significant **traction**. As nations grapple with the devastating consequences of climate change, the Association of Southeast Asian Nations and China have emerged as key players in fostering a green transformation. In 2021, ASEAN and China issued a joint statement on enhancing green and sustainable development cooperation and in 2022, both sides committed to strengthening common and sustainable development. Notably, 2021 and 2022 were designated as years of sustainable development cooperation between ASEAN and China.

2  In 2022, Cambodia, the **rotating** ASEAN Chair, proposed developing the ASEAN Green Deal to promote a sustainable and inclusive recovery from the COVID-19 pandemic. At the ASEAN Summit in Indonesia this month (May, 2023), ASEAN leaders adopted a Declaration on Developing a Regional Electric Vehicle Ecosystem as part of ASEAN's efforts to reduce greenhouse gas emissions, accelerate the transition  to clean energy, decarbonize the land transport sector in the region, achieve net-zero emission targets and improve energy security in each ASEAN member state and in the region.

3  It is crystal clear that a green economy is an important source of growth for ASEAN. Moreover, a green recovery and transformation will help ASEAN to achieve an economically and environmentally resilient future. According to a study by the Asian Development Bank, ASEAN's green recovery from the COVID-19 pandemic could potentially create $172 billion in investment opportunities annually and generate more than 30 million jobs by 2030.

4  As Southeast Asian economies are in transition toward more sustainable, resource-efficient and climate neutral economies, a regional framework on green recovery and green transformation is needed. In this connection, the development of the ASEAN Green Deal can serve as a regional framework for multi-stakeholder consultation and partnership building.

5  To realize the ASEAN Green Deal, five strategic policy interventions are recommended:

first, advancing a zero-carbon, affordable and resilient energy system; second, developing smart, low-carbon, water-secure and climate-resilient infrastructure and mobility; third, mobilizing public and private finance for green investment and transformation; fourth, accelerating and scaling up green innovation; fifth, implementing the United Nations Economic and Social Commission for Asia and the Pacific's Asia Pacific Green Deal and the Framework for Circular Economy for ASEAN Economic Community through the promotion of multi-stakeholder consultation and partnership building and the development of a green economic zone.

6   ASEAN and China <u>are endowed with</u> abundant resources and diverse ecosystems, providing a fertile ground for innovative green solutions. By <u>pooling their strengths</u>, the ASEAN members and China can tackle shared environmental challenges and drive sustainable development. Collaborative initiatives on renewable energy, environmental conservation and sustainable infrastructure will not only promote economic growth but also ensure the preservation of natural resources for future generations.

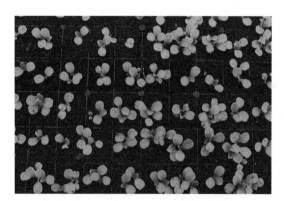 

7   Renewable energy stands at the forefront of the global transition toward a low-carbon future. Green investment is critical to realizing the low-carbon transition and a carbon-neutral economy. According to various studies, Southeast Asia needs about $3 trillion cumulative green investment to stay <u>on track</u> to reach 1.5 C target by 2030. In this connection, ASEAN and China can **leverage** their collective capabilities to <u>ramp up</u> investments in renewable energy sources such as solar, wind, hydro, and geothermal power.

8   China's growing influence in renewable energy development globally presents a significant opportunity for ASEAN to expedite its energy transition. This is particularly noteworthy considering the existing cooperation between China and ASEAN in various domains. To illustrate, China has invested approximately $31 billion in renewable energy projects in the ASEAN region from 2000 to 2020, **constituting** about 60 percent of the total foreign public investment in the region during that time frame.

9   Moreover, the involvement of the private sector is crucial to China's advancement in

renewable energy projects within the region. Starting from 2010, the private sector, acting as a sponsor and financing institution, has played a prominent role in China's engagement with renewable energy projects. In this regard, Chinese sponsors have pledged an investment of $3.6 billion in renewable energy projects as of 2023.

**10**  Electric vehicles present another promising avenue for cooperation between ASEAN and China. ASEAN is actively striving to establish favorable conditions within its domestic market to attract investments from EV manufacturers. In this context, China, being the foremost producer of EVs, holds significant importance. Strengthening partnerships with China through investments in EV products can generate positive spill-over effects, fostering innovation and growth in renewable energy-supporting industries across the region. This collaborative effort has the potential to drive the transition toward clean energy in the ASEAN region.

**11**  To facilitate a comprehensive green transformation, the members of ASEAN and China must foster a conducive policy environment. Harmonizing regulations, promoting knowledge exchange, and establishing regional standards and certifications are vital steps. By aligning their policy frameworks, these nations can incentivize sustainable practices, attract green investments and establish a level playing field for businesses operating in the region.

**12**  ASEAN and China's collaboration on a green transformation **transcends** regional boundaries and serves as an exemplary model for global cooperation to realize the necessary green transformation including the development and promotion of electric vehicles which China is leading.

**13**  The imperative for a green transformation is an **unparalleled** opportunity for ASEAN and China to lead the charge toward a sustainable and resilient future. By <u>capitalizing on</u> their strengths, fostering innovation and prioritizing cooperation, ASEAN and China can together embark on a common journey toward a greener tomorrow.

### New Words

**traction** /ˈtrækʃən/ *n.*
Traction is a particular form of power that makes a vehicle move. 牵引力
**rotate** /rəʊˈteɪt/ *v.*

If people or things rotate, or if someone rotates them, they take turns to do a particular job or serve a particular purpose. 使轮流；轮流

**leverage** /ˈliːvərɪdʒ/ *v.*

to use something that you already have in order to achieve something new or better  充分利用（资源、观点等）

**constitute** /ˈkɒnstɪˌtjuːt/ *v.*

If something constitutes a particular thing, it can be regarded as being that thing.  构成

**transcend** /trænˈsɛnd/ *v.*

Something that transcends normal limits or boundaries goes beyond them, because it is more significant than them.  超越

**unparalleled** /ʌnˈpærəˌlɛld/ *adj.*

If you describe something as unparalleled, you are emphasizing that it is, for example, bigger, better, or worse than anything else of its kind, or anything that has happened before.  无比的

## Phrases and Expressions

**grapple with:** to try to deal with or understand a difficult problem or subject  处理；解决

**be endowed with:** to naturally have a particular feature, quality, etc.  天生赋有，生来具有（某种特性、品质等）

**pool one's strengths:** to suggest that two or more people should come together and combine their respective skills and resources in order to reach a common goal or objective  集中力量

**on track:** doing the right thing in order to achieve a particular result  步入正轨；做法对头

**ramp up:** to increase or cause to increase; to increase the effort involved in a process  加强；增加

**capitalize on:** to use a situation to your own advantage  充分利用

## Exercises

**I Understanding the Text**

1. This text can be divided into four parts. The paragraph numbers of each part have been given to you in the following table. Now fill in the table with the correct choice from the following list.

① ASEAN and China's collaboration on a green transformation serves as an exemplary model for global cooperation.

② The introduction of ASEAN Green Deal.

③ Renewable energy and electric vehicles present two promising avenues for cooperation between ASEAN and China.

④ ASEAN-China cooperation to promote sustainable development.

| Parts | Paragraphs | Main ideas |
| --- | --- | --- |
| Part One | Para. 1 | |
| Part Two | Paras.2-6 | |
| Part Three | Paras.7-10 | |
| Part Four | Paras.11-13 | |

2. Focus on the parallel sentences. Pay attention to the sentence patterns and fill in key information.

① At the ASEAN Summit in Indonesia this month, ASEAN leaders adopted a Declaration on Developing a Regional Electric Vehicle Ecosystem as part of ASEAN's efforts to _____, _____, _____, _____ _____ in each ASEAN member state and in the region.

② As Southeast Asian economies are in transition toward more _____ economies, a regional framework on green recovery and green transformation is needed.

③ ...first, advancing _____; second, developing _____; third, mobilizing _____ for green investment and transformation; fourth, accelerating and scaling up _____.

④ _____ are vital steps.

**II Focusing on Language in Context**

1. Key Words & Expressions

A. Fill in the blanks with the words given below. Change the form where necessary. Each word can be used only once.

**rotate  leverage  constitute  transcend  unparalleled**

① The members of the club can _____ and one person can do all the preparation for the evening.

② Now some experts are proposing that we should take advantage of the teen brain's keen sensitivity to the presence of friends and _____ it to improve education.

③ Twelve months _____ one year.

④ From the great river to the sea, the river has fostered a number of thriving port cities that _____ centuries and remain prosperous.

⑤ The book has enjoyed a success _____ in recent publishing history.

B. Fill in the blanks with the phrases given below. Change the form where necessary.

**grapple with; be endowed with; pool strength; on track; capitalize on**

① The new government has yet to _____ the problem of air pollution.

② She _____ intelligence and wit.

③ What do you do to stay _____ with your eating, and healthy habits?

④ The Patriotic Education Law can help Chinese people carry forward their national spirit and _____ in the new era.

⑤ The rebels seem to be trying to _____ the public's discontent with the government.

2. Usage

A. Complete the following sentences in the text by using phrases with "as".

① _____, the Association of Southeast Asian Nations and China _____ in fostering a green transformation. （当各国努力应对气候变化带来的灾难性后果时；已然涌现成为关键角色）

② Notably, 2021 and 2022 _____ between ASEAN and China.（被定为可持续发展合作年）

③ _____, a regional framework on green recovery and green transformation is needed. （随着东南亚经济体正在向更可持续的、资源节约的和气候中性的经济体转型）

④ At the ASEAN Summit in Indonesia this month, ASEAN leaders , _____ _____, accelerate the transition to clean energy, decarbonize the land transport sector in the region, achieve net-zero emission targets and improve energy security in each ASEAN member state and in the region. （通过了《关于发展区域电动汽车生态系统的宣言》，作为东盟减少温室气体排放努力的部分举措）

⑤ In this connection, the development of the ASEAN Green Deal can _____ _____. （为利益攸关方进行多方协商和建立伙伴关系提供区域性框架）

⑥ Starting from 2010, the private sector, _____ _____ has played a prominent role in China's engagement with renewable energy projects. （作为发起方和融资机构）

B. Complete the following sentences in the text by using participles.

① In 2021, ASEAN and China issued a joint statement on enhancing green and sustainable development cooperation and in 2022, both sides _____ _____.（致力于共同发展和可持续发展）

② ASEAN and China are endowed with abundant resources and diverse ecosystems, _____ _____. （为创新性绿色解决措施提供肥沃的土壤）

③ This is particularly noteworthy _____ between China and ASEAN in various domains. （考虑到已有的合作）

④ To illustrate, China has invested approximately $31 billion in renewable energy projects in the ASEAN region from 2000 to 2020, _____ in the region during that time frame. （占对外公共投资总额的60%左右）

### III Translation

Translate the following passage into English.

中国是全球最大的煤炭生产国和消费国，煤炭占中国能源消费（energy consumption）的很大一部分。在未来，煤炭在中国总体能源消费中所占的份额将有所减少，但煤炭消费仍将继续呈绝对上升态势。中国今天面临着严峻的环境问题，而煤炭在造成空气污染方面起了很大作用。尽管中国的煤炭资源很丰富，但是我们应该开始寻找替代资源（substitute resources）。这样不仅能造福环境，从长远看也会换来经济回报。

### IV Pair Work

Talk about the measures the Chinese government has taken to reduce the carbon dioxide emissions from the perspectives of economy, energy, transportation, industry and policies.

### TEXT B

## Achieve Dual Carbon Goals in a Balanced Way

Li Wanxin

1   Energy in different forms, such as heat and electricity, fuels the economy and helps humans meet their basic needs. Although China has rich coal **reserves**, they are still finite in terms of quantity, and extracting coal is full of uncertainties, for it requires hard labor and has huge ecological and social implications. And this does not <u>agree with</u> the energy and resource conservation policy which China has been emphasizing on the road to modernization which emphasizes the harmony between humankind and nature.

2   True, fossil fuel-powered economic growth has <u>paved the way for</u> China to become the world's second-largest economy, but it has also brought large carbon emissions. That's why all local governments have energy-conservation and water-recycling departments to achieve self-sufficiency in energy.

3   The first binding energy intensity target was set for the 11th Five-Year Plan (2006-2010) period. The 13th Five-Year Plan (2016-2020) adopted a dual control policy for the total amount and energy consumption intensity, capping the total energy consumption at 5 billion tons of

standard coal equivalent annually. And energy consumption per unit of GDP (energy intensity) was expected to decline by 15 percent from 2015 to 2020.

4  On the global front, China rose to leadership position in the climate governance system when some other countries **shirked** their international responsibilities, with President Xi Jinping announcing at the 75th annual session of the United Nations General Assembly in 2020 that China will peak its carbon emissions before 2030 and realize carbon neutrality before 2060.

5  The 14th Five-Year Plan (2021-2025) for National and Economic Development and the Long-Range Objectives Through the Year 2035 started synchronizing the management of energy consumption and carbon emissions. It requires energy use and carbon intensity to decrease by 13.5 percent and 18 percent by 2025, respectively, compared to 2020. Overall, carbon intensity is expected to reduce by more than 65 percent by 2030 compared to 2005.

6  China is committed to meeting the medium-term decarbonization targets set for 2025 and 2030. The stakes are too high to fail and decisions made today will have far-reaching implications for the future. In fact, the country has learned the lessons of what <u>resorting to</u> last-minute measures for meeting the dual control energy consumption targets means.

7  From late September to October in 2021, many parts of China experienced <u>power outages</u>, <u>power rationing</u>, and severe disruptions in production, leading to market instability due to irregular supplies and rising prices of raw materials, which affected people's daily lives.

8  And then the severe drought in the summer of 2022 lowered the generation capacity of hydropower, making coal the **bedrock** of energy security. Coal is critical for grid stability, as it accounts for about 70 percent of peak load provision.

9  Since energy security, economic social stability, and carbon reduction are all desirable goals, they need to be synchronized. Dual control of total energy consumption and intensity mainly focuses on the power sector and large end energy users. But decarbonization demands a **holistic** approach involving energy consumption, diversification of energy source, structural upgrading of industries to enable them to move up the value chains, reducing the economy's dependence on heavy and construction industries, electrification of transportation and industry, modernizing agriculture, and ecological and land preservation, and restoration to increase carbon sinks.

10  Thus, dual control of total carbon emissions and energy intensity addresses the environment/climate-economy dichotomy. It has the potential to guide the formulation and implementation of a holistic and forward-looking long-term development strategy that is politically

viable, technically functional, administratively operable and financially feasible.

**11** The dual energy policy provides a good basis for aligning economic and environmental/climate objectives and stimulating collective action. For example, power generators, grid companies and local governments now have the incentive to work together to increase renewable energy generation and storage by overcoming challenges such as the curtailment of wind power and reluctance to share transmission infrastructure. It could also give rise to innovative measures to reform the power markets and **differentiate** customers by their preferences for energy mixes.

**12** Upgrading of traditional energy and carbon intensive industries such as steel, non-ferrous metals, oil refining, chemicals and building materials are already underway, and the efforts to reduce carbon emissions instead of direct energy consumption will provide more flexibility and minimize disruption in the industrial and transportation systems.

**13** China's new energy policy could help supply the missing links for the interconnected environmental, climate, economic and social issues, address the incompatibility in policy measures and help realize the dual carbon target of peaking emissions before 2030 and achieving carbon neutrality before 2060 via balanced, if not optimal, approaches.

**14** China has invested heavily in research and development for finding technical solutions to energy-related problems. The key challenge lies in building an all-encompassing carbon emissions **inventory** for economic and human activities and institutions and policies enabling collective action based on monitoring, reporting and verifying the volume of carbon emissions.

### New Words

**reserve** /rɪˈzɜːv/ *n.*

A reserve is a supply of something that is available for use when it is needed. 储备

**shirk** /ʃɜːk/ *v.*

If someone shirks their responsibility or duty, they do not do what they have a responsibility to do. 逃避

**bedrock** /ˈbɛdˌrɒk/ *n.*

The bedrock of something is the principles, ideas, or facts on which it is based. 基本原理，牢固基础

**holistic** /həʊˈlɪstɪk/ *adj.*

relating to or concerned with wholes or with complete systems rather than with the individual parts 整体的；全面的

**differentiate** /ˌdɪfəˈrɛnʃɪˌeɪt/ *v.*

If you differentiate between things or if you differentiate one thing from another, you recognize or show the difference between them. 区分

**inventory** /ˈɪnvəntrɪ/ *n.*

An inventory is a written list of all the objects in a particular place such as all the merchandise in a shop. 清单

## Phrases and Expressions

**agree with:** to have the same opinion as another person or to approve of something  认同；同意

**pave the way for:** to make it easier for (something to happen or someone to do something)  为……铺平道路

**resort to:** to do or use (something) especially because no other choices are possible  求助于；诉诸

**power outage:** the loss of the electrical power  停电

**power rationing:** Energy rationing primarily involves measures that are designed to force energy conservation as an alternative to price mechanisms in energy markets. 电力配给

## Exercises

**I Comprehension Check**

① Why is extracting coal full of uncertainties?

② What is the disadvantage of fossil fuel-powered economic growth?

③ What is the policy adopted in the 13th Five-Year Plan?

④ What is the dual carbon policy?

⑤ What is the advantage of China's new energy policy?

**II Translation**

① Energy in different forms, such as heat and electricity, fuels the economy and helps humans meet their basic needs.

② China rose to leadership position in the climate governance system when some other countries shirked their international responsibilities.

③ China has invested heavily in research and development for finding technical solutions to energy-related problems.

④ The dual energy policy provides a good basis for aligning economic and environmental/climate objectives and stimulating collective action.

⑤Thus, dual control of total carbon emissions and energy intensity addresses the environment/climate-economy dichotomy.

## TEXT C

### Climate Action to Build a Shared Future

Asit K. Biswas & Cecilia Tortajada

1  Speaking at the general debate of the 70th session of the United Nations General Assembly in September 2015, President Xi Jinping called for a new type of international relations **featuring** win-win cooperation and building a community with a shared future for mankind. The concept of "building a community with a shared future for mankind" was included in the report Xi delivered to the 19th National Congress of the Communist Party of China in 2017.

2   Since then, it has become one of the core concepts guiding the foreign policy of China.

3   In the report to the 19th CPC National Congress, Xi <u>elaborated on</u> his views on the shared future for humankind, saying it would lead to an "open, inclusive, clean and beautiful world that enjoys lasting peace, universal security and common prosperity".

4   Countries across the world are facing major problems which are too complex for any single country, no matter how powerful, to solve them alone. So all countries, big and small, should first agree to a common solution and then endeavour to achieve it and collectively contribute to realizing common prosperity.

5   Climate change is only one such global problem which is seriously **obstructing** efforts to build a community with a shared future for mankind. Therefore, every country must deliver agreed results —such as peaking their carbon emissions and achieving carbon neutrality—within a **stipulated** time, in order to address the common problem of climate change.

6   At the 15th UN Climate Change Conference in Copenhagen in 2009, China was one of some 120 countries that agreed to make efforts to keep global temperature rise to below 2 degrees Celsius. Since then, China has set a series of targets to be achieved within a specific period to transform the country into a green, low-carbon and circular economy. Specifically, China will work to meet the following targets.

7   First, China will reduce its energy consumption by 13.5 percent and carbon emissions by 18 percent per unit of GDP, both from the 2020 levels, while increasing non-fossil energy consumption to 20 percent and raise the forest cover to 24.1 percent by 2025.

8   Second, China will reduce carbon emissions per unit of GDP by more than 65 percent from the 2005 level and boost non-fossil energy consumption up to 25 percent, with 1,200 GW of solar and wind power generation, while increasing the forest cover to 25 percent by 2030. In other words, China will peak its carbon emissions before 2030.

9   And third, China would have <u>transformed into</u> a green and circular economy, with its energy efficiency matching that of the most advanced countries and non-fossil energy consumption exceeding 80 percent. This is to say China will realize carbon neutrality before 2060.

10   These are very ambitious targets by any standard.

11   China plans to limit the increase in coal consumption during the 14th Five-Year Plan

(2021-2025) period, while during the 15th Five-Year Plan (2026-2030) period, it will phase down coal consumption and peak its petroleum consumption. To achieve the climate targets, China has been vigorously developing renewable energy.

12    These ambitious goals are being **underpinned** by significantly strengthening basic research and promoting new cutting-edge technologies. Research and development will be intensified in low-carbon, zero-carbon and carbon-negative technologies, while major projects will be implemented to protect and restore all types of ecosystems, improve forest quality, and protect and restore wetlands and marine ecosystems.

13    Different State institutions have been developing macro plans to enable China to peak its carbon emissions before 2030 and achieve carbon neutrality before 2060. China has already made remarkable progress in many areas, pushing the country toward carbon neutrality, which will also help other countries to reach their respective goals in time if they use the cost-effective technologies developed by China.

14    But the International Energy Agency says that if the world is to reach net-zero by 2050, nearly 90 percent of energy will have to be generated using renewable resources, which can be made possible to a significant extent due to China's R&D efforts.

15    Take solar energy as an example to realize the progress China has made in fighting climate change. Some 15 years ago, Japan dominated this sector. But over the past decade, by investing heavily in R&D for solar and wind energy, China has become the global leader in this area. Also, primarily due to China's R & D efforts, global prices for generating per unit of solar and wind power have declined drastically—for instance, the cost of generating electricity through solar panels declined by an incredible 80 percent.

16    In 2021, China invested $381 billion in the clean energy sector—an **astounding** $146 billion more than the total investment in North America. And current estimates suggest that, through accelerated research, China will be able to reduce solar power generation cost to less than $0.03 per KWh by 2060.

17    In 2022, China had a solar power generating capacity of 393 GW, nearly one-third of the world's total, and it aims to increase it to 1,200 GW by 2030. Given its record, China could reach this target much ahead of schedule.

18    Yet to reach carbon neutrality before 2060, China has to increase its R & D budget

by trillions of dollars. The World Economic Forum estimates that China will have to spend $22 trillion on R&D during the 2020-60 period, but our estimate suggests the amount could exceed $30 trillion.

**19** The dynamics of the future of humankind will be determined by not any one single country but by the aggregation of all the countries' performances. In the final analysis, humankind has a common future. The world will prosper or **perish** together, though it is preferable to have a prosperous world with lasting social and economic well-being.

## New Words

**feature** /ˈfiːtʃə/ *v.*
have as a prominent attribute or aspect 以……为特点

**obstruct** /əbˈstrʌkt/ *v.*
obstruct progress or a process means to prevent it from happening properly 阻挠

**stipulate** /ˈstɪpjʊˌleɪt/ *v.*
If you stipulate a condition or stipulate that something must be done, you say clearly that it must be done. 规定; 明确要求

**underpin** /ˌʌndəˈpɪn/ *v.*
If one thing underpins another, it helps the other thing to continue or succeed by supporting and strengthening it. 支撑; 加固

**astound** /əˈstaʊnd/ *v.*
If something astounds you, you are very surprised by it. 使震惊

**perish** /ˈpɛrɪʃ/ *v.*
If people or animals perish, they die as a result of very harsh conditions or as the result of an accident. （因恶劣条件或事故）死亡

## Phrases and Expressions

**elaborate on:** to add more information or explain something that you have said 阐释

**transform into:** to change completely the appearance or character of something or someone, especially so that that thing or person is improved 转变

**phase down:** to remove or stop using something gradually or in stages 逐步减少

## Exercises

**Comprehension Check**

① What is the feature of a new type of international relations?

② What is one of the core concepts guiding the foreign policy of China?

③ What will be the outcomes of the shared future for mankind?

④ Why should countries first agree to a common solution?

⑤ At the 15th UN Climate Change Conference in Copenhagen in 2009, how many countries agreed to make efforts to keep global temperature rise to below 2 degrees Celsius?

## Supplementary Reading

TEXT A

### Coming Clean

Li Xinlei & Hao Junyi

1   China and the European Union have been cooperating in the field of clean energy for more than 30 years. Facing the challenge of achieving carbon neutrality, the cooperation has displayed some new patterns.

2   In March, the European Commission published two new acts: the Net-Zero Industry Act and the European Critical Raw Materials Act. The aim was to boost the EU's green economy and prevent the EU from deep reliance on other countries. In April, the European Parliament approved the legislation for the implementation of its Carbon Border Adjustment Mechanism, which is designed to reduce the effects of carbon leakage resulting from global trade in emissions-intensive goods, as part of the EU's "Fit for 55" package, which aims to align EU legislation with its objective of reducing net greenhouse gas emissions by at least 55 percent by 2030.

3   The recent actions of the EU have once again put the competition between the EU and China in the spotlight. Recently, China and the EU have been engaged in an increasingly sharp struggle for market share, material resources, and structural leadership. However, even though the increasing competition between China and the EU is genuine, it is in their interests to keep cooperating and coordinating with each other on clean energy. To make Europe the first climate-

neutral continent in the world, the EU released the European Green Deal in 2019 and wrote the goals set out in the Green Deal into the European Climate Law. China released its carbon neutrality goal for 2060 one year later in 2020. In the same year, it published its 1+N policy framework, consisting of a guiding policy and various supporting policies, to ensure the achievement of carbon peaking and carbon neutrality.

4  With both sides setting goals to realize carbon neutrality, China and the EU have rising demands for bilateral cooperation to achieve green growth. Recently, China and the EU increased the strategic importance of their partnership on climate change. On Sept 14,2020, the China-EU High-Level Environment and Climate Dialogue was initiated to forge green and digital partnerships. The mechanism provided another important institutionalized platform for China and the EU to strengthen communications and foster cooperation on climate change and clean energy.

5  Meanwhile, the interdependence of China and the EU in green trade has been increasing, which is also paving the way for their strengthened bilateral cooperation. There is an upward trend in the trade of products related to green energy between China and the EU, and the EU has become China's largest trading partner for exports of green products. According to statistics from Eurostat, in 2021, the EU imported solar panels (89 percent) worth 8.72 billion euros ($9.57 billion) and wind turbines (64 percent) worth 384 million euros from China, making China its largest import source of green energy products.

6  In addition, their complementarity in green industries highlighted by the energy crisis has built the base for strengthened Sino-EU cooperation. With its leading manufacturing ability for wind turbines and solar panels, China is the EU's largest source of extra-EU imports of wind turbines and solar panels and an important supplier of critical raw materials. In response to the global energy market disruption caused by the Ukraine conflict, the EU launched its REPowerEU Plan to make Europe independent from Russian fossil fuels well before 2030, setting out ambitious goals for its clean energy industry. To achieve those goals, the EU needs to strengthen cooperation with China.

7  To further strengthen cooperation between the EU and China, the two sides should continue promoting policy coordination, try to construct a practical and inclusive clean energy partnership network, and deepen cooperation on new agendas in multilateral climate governance.

8  First, scaling up global clean energy development to boost economic recovery and achieve climate neutrality goals requires clear international standards on green trade, green finance

and digital technology. A closer alignment of standards in those areas between China and the EU would provide a basis for the establishment of just global green economy standards.

9   Second, the existing China-EU climate energy partnership is expected to be further improved and expanded. While more interest conflicts in green infrastructure have emerged in recent years, there is great potential for China and the EU to expand their cooperation network by aligning their financial investment and technical support in third countries.

10   Third, besides bilateral policy coordination, Beijing can also pursue more coordinated action with the EU on new agendas within multilateral governance frameworks. For example, in the area of sustainable finance, China and the EU have initiated a working group on taxonomies under the International Platform on Sustainable Finance. In November 2021, with joint efforts from China and the EU, the IPSF Taxonomy Working Group published the first version of the Common Ground Taxonomy report, which provided a common classification tool for the global sustainable finance market.

TEXT B

## Chinese Energy Giant Strives for Carbon Neutrality, Launching Mega Carbon Capture Project

1   The China Energy Investment Corporation (China Energy) on Friday put into use a mega carbon capture, utilization and storage (CCUS) facility in one of its subsidiary coal-fired power plants in East China's Jiangsu Province, amid China's efforts to achieve carbon neutrality.

2   Attached to a generation unit at China Energy's Taizhou coal-fired power plant, the project will capture 500,000 tons of carbon dioxide every year, according to China Energy.

3   The facility has become Asia's largest CCUS project for the coal-fired power generation sector. It is also the world's third-largest, after one in the United States and another in Canada, China Energy said.

4   "During the project's trial run, the CCUS system demonstrated reliable performance and high safety standards, and the energy efficiency indicators and product quality are at or above their designed levels," said Ji Mingbin, president of China Energy's Jiangsu branch.

**Tech innovation**

5   At the CCUS facility, thick pipelines link the smokestacks with steeping towers. Flue gas would go through the towers. $CO_2$ from the emissions would bond with the amine chemicals at low temperatures in a tower, and the carbon would be released when re-heated in another. The separated $CO_2$, almost pure, is then pressurized and transferred to storage containers.

6   As its fourth and largest carbon capture project ever, China Energy has made substantial upgrades to its $CO_2$ capture and compressing technologies, largely reducing second-time energy consumption and cutting absorbent costs.

7   "It is not an easy process as we want to find an amine absorbent that is energy-efficient, highly stable and endurable," said Xu Dong, director of the carbon neutrality center of the Guodian Institute of New Energy Technology, which is affiliated with China Energy.

8   Unlike carbon capturing in other scenarios, the concentration of $CO_2$ from the flue gas is usually at or below 15 percent, Xu said, adding that the research team has completed more than 1,000 tests to find a suitable absorbent.

9   With self-developed amine mixtures, capturing one ton of $CO_2$ takes less than 90 kilowatt-hours of electricity, and the carbon capture rate is increased to over 90 percent and re-heating energy consumption is down by over 35 percent, China Energy said.

10   The team partnered with other institutes to produce the largest $CO_2$ compressor nationwide. "We have realized the 100 percent domestic production of equipment in our CCUS project," Xu said.

**Commercial operations**

11   Though the CCUS project is regarded as a viable way to realize carbon neutrality in the energy sector, it faces issues in making profits and in narrow application scenarios. "The inability to achieve the full consumption of $CO_2$ has been a blockage limiting the sustainable operation of CCUS projects," Ji said.

12   To tackle the problem, the China Energy Taizhou power generation plant undertook in-depth research in 2019, which showed that a vast number of chemical, ship-manufacturing and

food companies in Jiangsu had abundant market demand of approximately 1 million tons of $CO_2$ per year.

13  Ji revealed that the cost of producing a ton of $CO_2$ is around 250 yuan ($35), and all $CO_2$ produced and captured can be utilized, as the company has already secured contracts with eight firms. Primary applications for the captured $CO_2$ include dry-ice manufacturing and the production of shielding gases for welding.

14  "The purity of our $CO_2$ has reached 99.99 percent, which meets the requirements to be added to beverages like cola," Ji said, "In the future, we look forward to selling $CO_2$ to the nearby regions like Zhejiang and Shanghai, and maybe abroad to South Korea and Japan."

**Carbon reduction endeavors**

15  China lowered coal-fired power generation to below 60 percent of its total power generation in 2022. However, coal will remain as a primary power generation source for a long time, given the country's coal-dominated energy resource endowment.

16  China Energy, a coal-fired power generation giant, is one of the leading companies building pilot carbon capture and storage (CCS) projects in China. These projects are among the country's endeavors to achieve carbon neutrality by 2060.

17  The company launched its first CCS project of 100,000 tons in 2011, and a 150,000-ton CCS project in 2021. Early this year, it also realized the mineralization of $CO_2$ in a CCS project in Datong City, central China's Shanxi Province.

18  "China Energy will continue its endeavors to strengthen technological innovation, and contribute efforts to the country's carbon neutrality goal," said Zhang Changyan, a spokesperson for China Energy.

## TEXT C

### Forest Sinks Have Critical Role in Carbon Reduction

Li Hongyang

1  Top scientists have emphasized the importance of forest carbon sinks as having an irreplaceable role in reducing carbon in the atmosphere and ultimately combating climate change.

2  Yin Weilun, an academician of the Chinese Academy of Engineering and former principal of Beijing Forestry University, said that forests and grasslands hold a unique position in combating global climate change.

3  Forest carbon sinks are plants that absorb carbon dioxide from the atmosphere and fix it in vegetation or soil, thereby reducing concentration of the gas, Yin said.

4    Speaking on Tuesday at the Forestry and Grassland Carbon Sink Innovation International Forum, which is running parallel to the Zhongguancun Forum held in Beijing from Thursday to Tuesday, the scientist said that good forest management can increase carbon absorption capability.

5    Yin called for more research in forest management to ensure the sustainability of forests and their carbon sink capacity.

6    "The growth and development cycle of each tree species is different. For example, poplars reach their peak height in about 20 years. After that, they are unlikely to grow and the tree top will rot away."

7    To use land efficiently, forest managers must cut down some rotten trees and replace them with young ones so that the forest system can maintain its ability to reproduce and act as a carbon sink, he said.

8    Yin also stressed the need for increasing forest reserves, grasslands and wetlands to enhance their ability to absorb carbon.

9    At the forum, Du Xiangwan, former deputy head of the Chinese Academy of Engineering, said forests and grasslands can make multiple contributions to climate change mitigation as each cubic meter of wood growth can absorb an average of 1.83 metric tons of carbon dioxide, he said.

10    In addition, afforestation, grassland and wetland restoration, and desertification control not only protect biodiversity but also provide solutions for climate change adaptation and mitigation, he added.

11    Last year, the National Forestry and Grassland Administration set up a carbon sink research institute to study the country's potential for carbon sinks through forests and grasslands.

12　This research work will assess the spatial distribution of carbon sinks with the aim of understanding how to increase them, the administration said.

## Writing

Focus on one of the natural carbon sinks, such as forests, oceans and soil. Write about its mechanism to absorb carbon and how to protect it.

## Unit Project

微课

Discussion:

The carbon footprint is the amount of greenhouse gases (GHGs) that is released into the environment by an individual, a system, or an activity. Since there are various types of GHGs, for convenience, most carbon footprints are calculated by converting all GHGs to carbon dioxide equivalents. Please estimate your daily carbon footprints, and summarize what activities in your daily life have the largest carbon footprints. Discuss with your group and make a list of what alternatives are available to mitigate the carbon footprints in your community and thus help to achieve the dual carbon goals. Values of $CO_2$ emissions for typical daily activities are shown in Table 1. You can also use the Internet for additional information.

Table 1　The Carbon Footprint of Typical Daily Activities

| Daily Activities & Products | $CO_2$ Emissions (g) |
| --- | --- |
| Standard light bulb (100 watts, four hours) | 172 |
| Mobile phone use (195 minutes per day) | 189 |
| Washing machine (0.63kWh) | 275 |
| Electric oven (1.56 kWh) | 675 |
| Toilet roll (2-ply) | 1300 |
| Hot shower (10 mins) | 2000 |
| Medium car (1 km) | 192 |
| Bus (1 km) | 105 |

续表

| Daily Activities & Products | $CO_2$ Emissions (g) |
| --- | --- |
| Motorcycle (1 km) | 103 |
| Watching videos (1 h) | 36 |
| Weibo (30 min) | 93 |
| Using a computer (8 h) | 686 |